AF822503

Mechanisms and the Contingency of Social Causality

Mechanisms and the Contingency of Social Causality

By

Alexei Anisin

Cambridge Scholars Publishing

Mechanisms and the Contingency of Social Causality

By Alexei Anisin

This book first published 2022

Cambridge Scholars Publishing

Lady Stephenson Library, Newcastle upon Tyne, NE6 2PA, UK

British Library Cataloguing in Publication Data
A catalogue record for this book is available from the British Library

Copyright © 2022 by Alexei Anisin

All rights for this book reserved. No part of this book may be reproduced, stored in a retrieval system, or transmitted, in any form or by any means, electronic, mechanical, photocopying, recording or otherwise, without the prior permission of the copyright owner.

ISBN (10): 1-5275-8636-7
ISBN (13): 978-1-5275-8636-9

To Zoya

TABLE OF CONTENTS

CHAPTER ONE

INTRODUCTION

Hallways are filled with people wearing name tags of their titles and institutional affiliations. The smell of roasted coffee and sweet pastries circulates throughout the upper floor of an old building as a flurry of attendees wanders around trying to find their designated seminar room. Inside one of these seminar rooms, a large screen is illuminated by a projector displaying a set of images with equations, hypotheses statements, and an X Y plot. As the presenter concludes his argument, members of the audience nod and begin to raise their hands. "Interesting presentation and research question, but what's the mechanism?" ask Jane Doe and John Smith. This academic conference features typical presentations and typical follow up questions–which widely acknowledge that correlation usually does not mean causation. This is why the term mechanism is commonly used in academic discourse. Mechanisms imply causation and it is hard to come across an academic article without seeing the term mechanism being used either as a heuristic to supplement an argument or as some form of mediating force that is assumed to be operational in a causal relationship or correlation. Philosophers of science have extensively debated the utility of mechanisms in scientific explanations across disciplines throughout both the natural and social sciences (Norkus 2005).

Since the early 1990s, the use of the term mechanism has superseded the term "cause" in all English-written research papers by fifty percent. This effect is tenfold if compared to one century ago (Williamson 2011). Even earlier, for example from 1700 to 1900, the term mechanism started appearing in books and publications with great frequency–a trend that continued to not only increase across time, but bloomed into the 2000s alongside the rise of statistical applications and the data revolution. Figure 1 illustrates how the literary usage of the term mechanism transformed over the course of several centuries. This term went from being hardly mentioned in the early 1700s to being used at a much greater frequency by the 1900s and 2000s. These data are taken from Google Ngram viewer which is a publicly accessible tool that contains data on words from all books and texts written in the English language.

Figure 1.

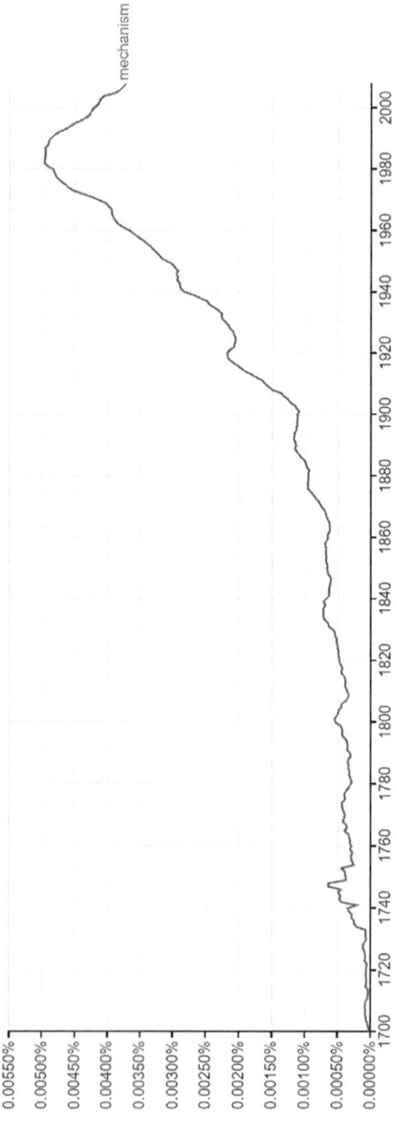

The origin of this term, like many others, stems to the Ancient Greeks. In English, German, Russian, and in Latin-based languages there is a Greek etymology of the mechanism (mekhanizm; mechanismus; mecanismo). Specifically, the contemporary usage of the term mechanism across dozens of languages has roots in the word *mechane* which was originally associated with the Ancient Greek idea of a deus ex machina; apò mēkhanês theós; God from the machine. This was a theistic plot changing tool or force that was used by authors of tragedies and comedies. From its outset, the mechanism was articulated to be a causal force that was a part of social outcomes and causality. Since then, there have been a number of semantic changes to the term, which over large periods of time, have resulted in semantic change. Civilizations have shifted from theistic to atheistic worldviews over the last two millennia, and in today's debates and discussions of mechanisms, philosophers of science and social science tend to regard mechanisms as forces that operate with varying degree of regularity in the empirical world.

Yet there is an intrinsic variance in how mechanisms are conceptualized and defined. This is widely acknowledged by scholars. Although most debates have centered around whether mechanisms can either supplement or attempt to supplement strict laws of nature as well as deductive-nomological explanation (Leuridan 2010), defining mechanisms has been problematic. Mahoney (2001) notably identified nearly two dozen of different definitions of mechanisms across disciplines. Mechanisms took on the following roles: variables that explain correlations; as forces that are within a "black box" that lays between independent and dependent variables; as "mid-level" theories that can account for different empirical phenomena under recognizable patterns or under unknown conditions; as an unobserved entity that generates a given outcome of interest (Mahoney 2001, 581). More than 20 years have passed since Mahoney's overview, and one would probably not be wrong if they were to argue that more definitions have since arisen, and greater debates have since ensued.

Similarly, decades ago, Elster issued a "plea for mechanisms" in light of prediction shortcomings (Elster 1989; 1998). Little (2012) has argued that mechanistic explanations of the social world are the best explanations we can possibly both hope for or expect (Little 2012). In scholarship on the origin of mechanisms, focus has been placed on linking mechanisms to the historical development of mechanical philosophy. Glennan and Illari (2018) identify mechanistic explanations and mechanical philosophy to date back to Democritius (460–370 B.C.), and Popa (2018) and Roux (2018) similarly trace the prehistory of the idea of a mechanism to

Democritus, the Epicureans, Aristotle, and Descartes. In these accounts, the idea of a mechanism is linked to early (and some of the first) historical instances in which mechanical philosophy (and ideas of mechanics) were formulated. While there is nothing particularly erroneous in these approaches, as this book will demonstrate, such approaches do not actually touch on the primary basis that lays behind the emergence of a mechanism–a basis that is metaphysical in its roots and was not limited to any particular philosophical framework or school of thought. In asking what the ontological source of a mechanism is comprised of, this book explores the etymology of this term and the socio-historical contexts in which it emerged.

Underlying the aforementioned discussions is something much grander that hitherto has not been acknowledged by philosophers of science, social scientists, and research methodologists. The term mechanism has historically functioned as a construct of social causality. This does not mean the mechanism is in anyway an idealist construct or something that does not exist independent of discourse. Rather, this book demonstrates that the origin of this term stems to human beings' attempts to rationalize aspects of the noumenal world [ideas of God(s) and spirit], through usage of synthetic a-priori abilities and qualities. This is how the term mechanism first arose. These processes of rationalization were initially applied to social causality by the Ancient Greeks to explain outcomes in the phenomenal world. The creation of the *mechane* came into being when a wooden mechanical device was designed for Greek theaters in Athens to holster up characters from the ground and ascend them into the sky to illustrate the deus ex machina in playwrights. The mechanism functioned to deploy an event in a story during theater shows (Papadogiannis, Tsakoumaki and Chondros 2010). A lever that was high-above stage would swing down to bring characters directly onstage or through a hidden trapped door (Ashby 1998). Historians describe the first mechane as comprising a system of a winch and pulleys containing a wooden beam with an attached harness that could both rise and swivel (Storey and Allan 2014, 45).

Populations in these ancient eras of history attributed causal forces to the works of God(s). An intriguing biblical utilization of the mechanism can be found in the parable of a large fish swallowing a prophet (Jonah) during a storm. The fish then saves him through delivering him to Nineveh, an ancient city in Upper Mesopotamia (Bell 2013). As time went on, the gradual decline of religion in social life (and the concurrent rise in science), led to posterior interpretations of mechanisms that were not tied to the noumenal world in any purposeful way. Into the modern era of history, the mechanism became a posteriori concept based on empirical knowledge. It became a much more technical and positivistic term. The mechanism thus has constituted both discursive and material realities and has represented ideas from both the noumenal and phenomenal worlds. Throughout this book, I situate the term mechanism and its representational meaning(s) more so with understandings of social causation rather than with natural causation. The reason for this is that societal tendencies and worldviews tend to be inclusive of understandings of causation in the natural world. As such, I assume that the term social causality is synonymous with social causation and that the mechanism has historically functioned as the most integral element or component of social causality.

The idea of social causation is not straightforward or easy by any means. Kincaid (2009) argues against methodological individualism and reductionist approaches that are commonly deployed in the social sciences by stating that social causation arises from both structural entities and individual agents. I agree with this important distinction, but also situate my treatment of social causation along with discussions of natural causation–both of which are ultimately apart of the what Kant considered to be the phenomenal world or realm. This brings us to a major theme of this book. Throughout every single era of human history in the Western world (broadly conceived) ranging from Ancient Greece to our contemporary civilization, common societal-level understandings of social causation have always been coherent and constitutive of a shared apprehension of the world. There are macro trends that are intrinsic to different periods of human civilization which are reflected in discourse as well as culture. Throughout this book, evidence is presented to demonstrate that the term mechanism and its etymology as well as historical semantic change(s) are very much tied to such common understandings of social causality. The term mechanism has historically enabled populations to categorize how they perceive causal forces operate in the social world.

While there are dozens of definitions of the term mechanism, few have considered what this term actually symbolizes and means from a grander perspective. In this sense, this book's analyses echo a Heideggerian approach to philosophy (especially his works on technology) because I ask what the essence of a mechanism is? As a term in discourse and as a term that has taken on different meanings throughout different historical eras, what is a mechanism? Over the last two millennia, epochal differences can be observed in how mechanisms were interpreted and utilized in discourse and culture. To date, no work has examined these changes nor has any work specifically addressed the etymology of the mechanism. Our lack of understanding pertaining to these topics reflects significant gaps in knowledge because the term mechanism constitutes arguably the most direct linguistic representation of human beings' understanding of causal forces. This book carries out an analysis of heterogeneous different eras of history to identify the origin of the mechanism and to trace the semantic alterations that this term has gone through. Through five chapters (which form the basis of the text) along with a seventh chapter (the conclusion) I discover that semantic alterations associated with the term mechanism can be observed directly with changes in referents (physical changes in the world) and changes in categorization (forms that associated with worldviews).

Prior to the onset of modernity, understandings of mechanisms were remarkably different because they were tied to either a polytheistic (Ancient Greek) or theistic (Christian) understanding of the social world, both of which were religious in character. Specifically, the mechanism was representative of forces that were God-like. Changes emerged in the Roman empire and early Christianity where the mechanism remained a theistic force but became epideictic of one deity, rather than a heterogeneous collection of deities. During the Middle Ages and into the era of modernity, notable novelists and writers utilized the mechanism as a metaphysical plot shifting force. However, with the era of modernity, major alterations arose in terms of how the world was categorized and this paved way for the scientific usage of the mechanism. At the time of the writing of this book, the most recent changes in the etymological and semantic characteristics of the mechanism (and in turn, social causation) are directly tied to the digital revolution and the emergence of artificial intelligence (AI). The emergence of AI, it appears, has rendered a completely new understanding of mechanistic activities and social causation due to manifestations of autonomous and non-human mechanistic actions and forces.

Theoretical Framework

This book explores these cross-historical dynamics through tracing the broad etymology of the mechanism and offers both scholars and members of the general public a different glance into the underpinning of social causality and a new innovative assessment of the contingency of social causation. Throughout the chapters of this book, insights are drawn from a variety of different thinkers, philosophers, scientists, and regular people that contributed to historical innovations. Each chapter attempts to map out the overarching macro tendencies inherent to socio-cultural configurations that populations lived within. This approach makes this book accessible to both students and scholars, and different chapters should be drawn upon by different audiences accordingly. Moreover, the analyses presented in this book are complimented by explanations of the different philosophical and scientific contributions that groups and individuals put forward to what truly has been an evolutionary process of knowledge advancement that began thousands of years ago and continued up until the writing of this book.

It is necessary to differentiate this book from the rather large wave of anti-positivist thought and output where mechanistic explanations have been rejected. The emergence of post-modern, feminist, post-structuralist, post-colonial, and other critical traditions in social sciences and humanities have led to arguments against mechanistic approaches. My intention in this book is not to take sides with any particular school of thought or scholarly tradition. The aim of this book is to fill a major gap in knowledge through a philosophical approach that draws from specific aspects of different philosophical frameworks. While this book mentions many different thinkers and discoveries of the last two millennia, the following three figures are especially important to the different frameworks I draw on to make sense of grandiose historical eras and processes that transpired within and in some cases, throughout these eras. The first figure is Nikolai Berdyaev (1974-1948) who was a Slavophil philosopher that was exiled from the Russian Empire by the Bolsheviks and spent the remainder of his life in Western Europe. Berdyaev is considered to have been a highly original existentialist philosopher who many also refer to as a philosopher of spirituality and personalism. A particular work of Berdyaev (1952) titled, *The Realm of Spirit and the Realm of Caesar*, is drawn on and its particular categorization of the relationship that human beings have to nature and the cosmos. There are four of such periods that arose in history (a fifth was projected by Berdyaev to arise in the future). We currently are living in the fourth. This categorization helps to make sense of the changing trajectories that humans have lived through including polytheism, paganism, Christianity, modernity, science, and the advent of computational and digital realms. I also draw from Berdyaev's existentialist assumptions about personality and subjectivity.

The second major thinker whose ideas are frequently drawn on throughout this book is Immanuel Kant (1724-1804). Perhaps the most influential and important philosopher since Aristotle, Kant's contributions to philosophy are so significant that an entire encyclopedia can be written on them. For my purposes, I draw on Kant's ideas he put forward in his *Critique of Pure Reason* (1781) among other works. Kant's ([1797]; 1997) phenomenal and noumenal distinction is frequently utilized throughout different chapters of this book to describe how knowledge was attained on mechanisms in different eras of history. Importantly however, Kant's ideas on the phenomenal and noumenal realms are, as subsequent chapters of this book will reveal, incomplete. Although I assume that the phenomenal world exists and that the noumenal world also exists (but is not accessible through rationalistic categories of understanding), there is more to this book's framework than the distinction between phenomena and noumena.

One may identify potential inconsistencies with combining ideas of Berdyaev and Kant as the former was not in full agreement with the latter. Both figures were strong advocates of freedom, yet Berdyaev's faith played a much more prominent role in shaping not only his life, but his decisions in life such as accepting the Bolsheviks' proposal to send him away on the famous "philosophers ship" that purged dozens of intellectuals from the newly forged communist regime in the early 1920s. It is impossible to read a chapter of any of Berdyaev's books and not see a reference to Kant and some component of his thought. A particular difference between the two is that Berdyaev has been widely labeled as a Christian existentialist because his Orthodox faith underpinned the formation of nearly all of his arguments on philosophical topics. Perhaps the most salient difference is in Berydaev's ontological assumption of the human condition and how it relates to the noumenal realm. Berdyaev argued that, "the very fact of the existence of man is a break in the natural world and proves that nature cannot be self-sufficient but rests upon a supernatural reality" (Berdyaev 1937, 46). Only through human beings can spirit be revealed, argued Berdyaev, and this is the highest value of man. Along these lines, Berdyaev argued that being (if interpreted in an objective manner) does not have primacy over mankind, but rather, mankind has primacy over being because it is revealed "only in and through man" (Berdyaev 1937, 7). While Berdyaev did not take issue with Kant's important arguments on the limits of our knowledge, he did nevertheless believe knowledge is spiritual in its preconditions. This makes Berdyaev's position very starkly opposed to a Kantian transcendental consciousness.

Further, Berdyaev argued that the noumenal realm was eschatological and existential, and it is within this realm where God and theism subsist. According to Kant, the categories of our mind cannot adequately grasp metaphysical issues including theism (God), but we can identify the spatial and temporal characteristics of the phenomenal world through our synthetic a priori cognitive capabilities. This made Kant's reasoning kin to both the empiricism and the rationalist waves of his era. Importantly however, when considering Berdyaev's position, emphasis must be placed on his claim that, "the a priori forms which are supposed to vindicate the validity of knowledge have no direct relation to the concrete man who is the knower" (Berdyaev 1937, 10). Echoing Dostoevsky, Berdyaev believed humans were much more capable of obtaining and reaching spiritual heights through creative activities rather than through accepting the Kantian idealist postulate that a-priori transcendental consciousness is the end all. For my purposes throughout this book, I will assume there exist both phenomenal and noumenal worlds, and that God and theistic forces subsist within the latter, while materialism and causality exist within the former. The fascinating aspect about the origin of the term mechanism is that it constitutes an attempt by human beings to account for social causality through constructions and articulations that stem to both the phenomenal and noumenal worlds.

Mechanisms Across Phenomena and Noumena

Even with my adoption of the Kantian phenomenal and noumenal distinction, I do not agree with one of Kant's major points of orientation. Zizek (2012) refers to Kant's framework as being riddled by a "hidden arrogance" because of its assumptions about our restrictions to finite understanding. Since knowledge is limited in Kant's philosophy, we cannot ever know the totality of the universe. Because of this constraint, Zizek argues that the Kantian framework actually continues to represent this infinite task as a task that some other infinite understanding would be able to accomplish, "as if the problem is simply one of extending or extrapolating our capacity to infinity, rather than changing it qualitatively" (Zizek 2012, 211). As I have already noted, the implication I associate with the noumenal realm is not one that necessitates a "naturalist-determinist" idea that Zizek associates with Kant's conceptualization of noumena. Nor do noumena represent what Zizek refers to as self-limitation or pure negativity that are inaccessible to our experience.

Along these lines, a word must also be put in about Hegel, as I anticipate that some will consider the line of argumentation put forward in this book to be somehow related to a Hegelian framework and his grander philosophy of history. Indeed, Hegel shares much with Kant but he also was highly critical of him. As Priest argues, "in these philosophers, for the first time, we arrive at a general recognition of the contradictory nature of the limits of thought, together with a theorisation of how and why this occurs" (Priest 1995, 81). Yet even with their similarities, Hegel's criticisms of Kant are widely acknowledged to pertain to the "bifurcated world" and a capitulation to skepticism in Kant's philosophy (Berthold-Bond 1989, 43). The noumenal realm is an empty world, argued Hegel. As noted by Berthold-Bond (1989), "the central epistemological issue between Hegel and Kant has to do with the alteration of the object by consciousness in its activity of thinking that object" (Berthold-Bond 1989, 47). In Kant's philosophy, the altering activity is an "ultimate barrier" while for Hegel the recasting of the object by consciousness allows the object to appear in its "true light" (Berthold-Bond 1989, 47). In this sense, Hegel believed we can go into or behind the noumenal realm and to uncover things that can be seen in what he referred to as a native realm of truth (Solomon 1983, 425). Some have associated these ideas with Hegel's conceptualization of the unconsciousness. Hegel (2018) believed that a super-sensible (noumenal) realm does not exist as a separate entity because it is mirrored within the phenomenal realm. History is thus a reality of God and our ideas are a manifestation of God's ambition or will. Zizek, an earnest Hegelian, summarizes this logic as follows, "as Hegel put it with unsurpassable clarity in his Phenomenology: behind the curtain of phenomena, there is only what we put there" (Zizek 2012, 282).

Zizek's popular synthesis of Hegel's framework implies that the internal limitations of reality are in a gap that exists somewhere between phenomena and noumena. This is how Hegel was able to identify what Zizek believes to be a radical form of negativity that is within the subject itself as a form of correlative subjectivity which reflects the ontological incompleteness of phenomenal reality. Further, in Zizek's (2012) argument against the phenomenal and noumenal distinction, it is noted that, "there is no mysterious gap separating us from the unknown, the unknown is simply unknown, indifferent to being-known. In other words, we should never forget that what we know (as phenomena) is not separated from things-in-themselves by a dividing line, but is constitutive of them: phenomena do not form a special ontological domain, they are simply part of reality" (Zizek 2012, 211). Echoing Hegel, Zizek argues that the gap between proposed phenomena and noumena is "purely cognitive" through referencing research on neuronal processes in the brain. While I agree with Hegel's claim that perceptions of "real" phenomena can be proven through identifying their logical basis or rationalized parameters, I still adhere to the assumption that a noumenal realm exists.

It would be erroneous to believe that human thought can attain knowledge of the entire scope of subjectivity. The presuppositions that are voiced by both Hegel and Zizek reflect both thinkers' entrenchment in objectification. Berdyaev argued that Kant approached the problems associated with objectification in identifying some of the limits to knowledge in the world of appearances, but did not go far enough because this was not the "true" world (Calian 1969, 116). In this sense, Berdyaev's view of the noumenal realm as being both subjective and eschatological is accommodating. Berdyaev argued that humans are in a struggle against objectification and that his own framework was a spiritual revolt of noumena against phenomena (Calian 1969, 117). The argumentation put forward in this book is much closer to Berdyaev's conception as I believe that the noumenal realm reasserts mystery into our existence which has hitherto tended to get lost in an objectified and rationalistic world, especially in the context of "modern" post-Kantian philosophy. In this respect, even though this book draws from Kant's distinction between phenomena and noumena, it is rooted in a larger framework that is neither fully Kantian nor Hegelian.

Our senses and our perceptions can produce objective measures of the phenomena, but we will never uncover the full scope of characteristics about the noumenal realm because it is marked by spirituality that can only be approached metaphysically and existentially. This is why Berdyaev argued that knowledge of noumena is through intuition and not sensory input (Collier 20014, 187), and also why he warned against scientifically ascribing noumena to the properties of phenomena (Obolevitch 2019, 122). In my view, objectifying or attempting to shoot down the noumenal realm is an erroneous epistemological maneuver. Leaving the noumenal realm to spirituality and existentialism is the only way to account for the possibility of putting forward a framework (even if limited in its ambition or scope) that will withstand the test of time and historical developments.

From early periods of history, into the Middle Ages, then leading into the modern era (rise of science), and the contemporary digital age, socio-cultural configurations and tendencies have intersected between both phenomenal and noumenal realms. Early societies put forward attempts of rationalizing the noumenal realm through a-priori synthetic knowledge (which Kant argued is a basic tenancy of human thought) and it was a particular attempt in Ancient Greece that led to the creation of the mechanism. The Ancient Greeks had very little experimental knowledge, they did not yet conquer the natural world, and were submerged within cosmic, polytheistic worldviews. The mechanism was created as part of a process of rationalizing a polytheistic cosmos. In other words, components of the noumenal realm that were prevalent throughout the ideas and socio-cultural tendencies of Ancient Greek societies were physically transmigrated into the creation of a mechane (mechanism). As time went on, populations shifted away from articulating ideas in their cultures and understandings of social causality through theistic prisms. By the rise of the modern era, theism declined, more emphasis started being placed on the empirical (phenomenal) world. This tendency was exacerbated and excelled in the mid twentieth century and eventually turned into a morphed into our current point in history–the digital age. With these shifts and vast historical transmutations, semantic changes associated with the mechanism can be observed as can understandings of social causality. Below, the different historical time periods that this book investigates are split into sections to offer readers a preview.

Linguistic and Semantic Change

Generally, linguistic scholars agree that semantic changes of words do occur over time, especially when viewed over multiple centuries. There is regularity inherent to semantic change, and some linguists argue that terms are even predictable in how they will change (Traugott and Dasher 2001). The Ancient Greek conception of a mechanistic force was contingent on the polytheistic world views of its era. During this time period, humans were still submerged into cosmic life and their relation towards nature was based upon magic and myth (Berdyaev 1952, 47). It is in this particular milieu that the present investigation of semantic change of the term mechanism has to begin. Broadly speaking, semantic change of terms has been observed to be brought about by a heterogeneous collection of factors including those that are linguistic, psychological, socio-cultural, or encyclopedic in origin (Blank 1999). Grzega (2004) includes over a dozen different motivations that are associated with why semantic change arises–some of which are attributed to language disguising (misnomers), difficulties in attributing correct meanings to denotatives, social reasons, institutional and legal factors, morphological reasons, cultural salience, and most crucially, changes in referents and categorization. Changes in referents entail that there are physical changes in the world, whereas changes in categorization mean that changes in world view occurred. Semantic changes associated with the mechanism have come alongside physical changes in the world and simultaneously, with changes in worldview. In Ancient Greece, mechanistic forces were mythical, cosmic, and theistic. They functioned as plot changing agents in the many tragedies, comedies, and masterpieces of early literary works and theater. The effect of mechanistic actions in theater were tied to various deities of Ancient Greek religion (polytheism). Notable works included *Clouds* (Aristophanes), *Medea* (Euripides), *Thesmophoriazusae* (Arisophanes), *Orestes* (Euripides), Elektra (Sophocles), *Oresteia* (Aeschylus) among others drew upon the deus ex machina to solve their plots or to drive the entire purpose of a given story. In the public sphere, these massively important cultural stories were portrayed via enactments in theater and were watched by thousands of attendees (archaeologists have even projected that some theaters contained more than 12,000 seats). Story lines were also re-told for centuries after they were written.

While the pre-Christian era was dominated by the belief in and worship of multiple (in some contexts, many) gods (Hellenis, Pagans), the decline of Greece and rise of Rome coincided with an eventual shift away from polytheism to one God (theism). The ascension of Orthodox and Catholic forms of Christianity were salient as the belief of Jesus Christ became widespread in the Western world and into the Byzantine empire. With these changes in worldview, the idea, usage and symbolic interpretation of the mechane made its way into Rome and here the concept remained popular. Crowds continued to witness the mechane and would react with astonishment and ponder over the gods (Cunningham 1954). Roman theater was alike to Greek theater, and the usage of mechanes on stage remained prominent. In addition, the Romans used a mechane in funeral settings. The mechane went from being bound to forces that connected the populace with deities to eventually only being illustrative of a single deity. This occurred during rather turbulent and dictatorial times in Rome, and the shift was specifically solidified in the centuries after Julius Caesar's death.

The impact of this shift in belief systems cannot be understated as this was a major change in world view. The Church merged itself with the polity, making the emperor of the Holy Roman Empire an all-powerful deity. This set the stage for a new objective reality for much of the world–a reality that Berdyaev refers to as the "Realm of Caesar." Technological advancement was limited withal, and nature was still a hierarchical order established by a theistic power (Berdyaev 1952, 48). Hence, the mechanism remained theistic in its characteristics. Furthermore, several centuries of relative peace (Pax Romana) ensued under the sovereign Augustus and rulers after him, while religious conflicts occurred as pagans clashed with Christians until eventually (sometime in the 4th or 5th century AD) Christianity reigned paramount over paganism. Ignatius of Antioch was a disciple of John the Apostle and his articulation of the mechane can be traced in letters and teachings about Jesus Christ throughout the empire and the Middle East. As many recognize, biblical scriptures state that Jesus Christ was crucified on a Cross. The Cross for Ignatius of Antioch is a cosmic hoist and mechane that propels people to heaven. He wrote, "O Cross, thou hoist to heaven! The Cross was driven into the ground – and behold, idol worship was destroyed. No ordinary wood is this, but the wood that God used for victory" (Cardinal Ratzinger 2016). Ignatius' articulation entails that the mechane does not cast down God into the masses but that the Cross raises people from their lowest level to Jesus Christ and in turn, God. In this sense, common people can be raised out of paganism and into spiritual connection with one deity–God (at one point, the term Pagan literally meant non-urban peoples of the empire). Moreover, in Ignatius' articulation, the Cross of Jesus Christ is the figurative mechane, whereas the rope to the mechane is the Holy Spirit (Ballard and Holmes 2006, 9).

The association between the mechanism and a theistic power remained intact throughout the existence of the Roman empire. It was then carried over into the Middle Ages and was most significant in the late period of this era as observed in outstanding works of literature.

Ignatius' articulation may be even more significant than one would assume because it signified a reversal shift that enabled the mechane to be cognitively conceptualized by the masses. By the 14th century, Italian architects improved many aspects of theater engineering. For example, the *Paradiso* was a unique system of ropes hoisted by a machine that would elevate angels and other characters above stage (Brockett et al. 2007). Even with technological advances in areas such as theater and in urban life more broadly, populations still conceived and interpreted phenomena via a religious basis in daily life. Humans were still bound to mother-earth and were living in a telluric epoch (Berdyaev 1952, 48). By the time the era of the enlightenment came around, novelists, play writers, and other literary contributors already started to shift away from conceptualizing the mechanism as a God-resembling force to simply incorporating external plot-changing forces that were not necessarily theistic.

Numerous examples can be observed in literature from the high Middle Ages in which mechanistic plot solutions were deployed but were not necessarily theistic in their character as writers still relied upon the mechanism as an articulation of a forceful action and event, but did not link it to theistic powers. Problems in a given plot were solved via the mechanism and when this transpired, the reader or viewer would be forced into a major dilemma as the structure, cohesiveness and transcendent worth of a given plot would either fall into question or be resolved. The mechanism remained influential into the great classics of literature and theater in works such as Shakespeare's *As You Like It* or in *Pericles, Prince of Tyre*, in Tolkien's *The Lord of the Rings* and *The Hobbit*, in Dickens' *Oliver Twist*, in Golding's *Lord of the Flies*, in H.G. Wells' *The War of the Worlds*, among others.

With the onset of modernity, populations in much of the Western world experienced a decline in religious beliefs. While this slow-moving process did not unfold in one year, ten years, or even one hundred years, it did open up space for what was truly a radical change in human thought. The era of the enlightenment, the onset of modernity, and the emergence of liberalism and industrialization contributed to masses perceiving themselves as being free from God's antecedent oversight of their lives. This paved way for both separation between church and state as well as the general mechanization (human control) of nature. The mechanization of nature entailed those human beings, for the first time, garnered control over natural forces. Nature ceased being a natural hierarchical order controlled by theistic forces. During this era, science and the scientific method developed and in time, became paramount in their influence on society. Nietzsche's philosophy epitomized these changes, especially in his idea of *Gott ist tot* (the Christian God is dead).

The solidification of the principle of reason (as articulated by Leibniz in the 17-18th centuries) was significant in its consequences. This principle entails that everything in the world must have some reason or cause behind it. For instance, if there is a given entity (z) then there has to be an explanation for why it exists. The same exact logic is applied to every given event and every given proposition. For the first time in history, natural phenomena could be explained and, in some cases, predicted by physical laws. For example, the *Law of the conservation of mass* was discovered in the eighteenth century and states that in any closed system, a given mass of substances that are produced by a chemical reaction are always the same as the reactant. In other words, in any given chemical reaction, mass is unvarying and atoms do not disappear, say when say a piece of wood burns into ashes. Instead, atoms get reshuffled and reconfigured. As many other findings, this discovery helped change the status and position that science had over daily life and society.

Since nature was no longer dictated by theistic forces, understandings of causality changed greatly. No longer were causal forces in nature and daily life linked to theistic powers, but rather, the rise of scientific inquiry enabled widespread cognition to emerge about properties and material of the world. It was here where mechanisms were assumed to be operative as forces that could be discovered by human beings. Crucially, causal forces were not tied to polytheism or theism. This brings me to the primary intention behind why I wrote this book. While hundreds of studies and many different chapters in edited monograph volumes continue to be published surrounding the usage of mechanisms, attempts at defining them, and implementing mechanistic explanations within causal and scientific-based social inquiries, there currently is a grandiose lack of understanding about the contingency of this term and the relation that it has to social causality. This book takes what to my knowledge is the first step at fulfilling a cross-historical overview from Ancient Greek Theater to AI and the digital era of communication.

Causal Knowledge

Indeed, this book's investigation of social causality and semantic changes associated with the term mechanism necessitates that these concepts are made intelligible and defined with relation to previous frameworks. To understand what causal knowledge means, we first must consider Aristotle's Four Causes which has arguably formed the basis of Western philosophical thought and has underpinned the ideas of many philosophers and scientists' understandings of causality. Falcon (2022) accurately describes Aristotle's framework on causality as "causal pluralism," as it features his famous interrelated Four Causes, each of which contain a clear and distinct kind of cause. The four causes originate to Aristotle's explorations of the natural world and form the basis of what he argued must be in place for causal knowledge to arise. The four types of causation include 1) the material cause (which is the material substance out of which a cause arises); 2) the formal cause (which is the form that the cause will take on or will arise from a cause); 3) the efficient cause (which is the primary type of change associated with the cause); 4) the final cause (the end product or purpose of the cause) (Aristotle 1966). In articulating these different types of causation, Aristotle was particularly interested in explaining causal knowledge as applied to the natural world which is why Falcon (2022) makes the crucial point that his framework was teleological and did not depend on the application of psychological concepts (desires, beliefs, intentions, etc.). Many attribute Aristotle's framework of causes to the creation of the first bonafide deductive approach to study the natural world (and in turn, one of the first scientific frameworks put forward by any human being). Gerring (2001) argues that the third of Aristotle's causes, the efficient cause, has received the most attention in modern discussions because it is commonly be associated with an agent that produces change in something else. This pertains to a very bold (and in my view, correct) observation in that definitions of the word "cause" in English dictionaries tend to revolve around the notion that a cause is an agent because it "generates, creates, produces, effects" (Gerring 2001, 133). What then, sets my exploration of the term mechanism apart from social causality, and how does the aforementioned definition of the term "cause" relate to Aristotle's four causes and current definitions of the term "cause?" Although these are quite dense questions, they must be addressed in this introduction because statements put forward in subsequent chapters run risk of being difficult to interpret for the reader. First, as outlined in previous paragraphs, this book's analyses rest on the assumption that the noumenal and phenomenal worlds are distinct from one another. Aristotle's four causes, importantly, are akin to the phenomenal world.

This leads us to the next critical point, mechanisms are not necessarily tied to either of these realms, and in different periods of history, the understanding of what constitutes a mechanism has been influenced by ideas stemming between both worlds. For example, the original "mechane" was a physical construction that was used in theater, while the idea of the deus ex machina was a metaphoric or symbolic discursive construct that led populaces to believe social outcomes were determined by the gods. As subsequent chapters of this book will reveal, early history was marked by populations' attempts of rationalizing the noumenal world and explaining social outcomes through such processes of rationalization. This sets apart the historical content of the term mechanism from the modern term "cause," because the former neither inclusive to either the phenomenal or noumenal worlds while the latter is inclusive to the phenomenal world. As such, my assumptions of what constitutes social causality reflect what I commonly refer to as the populational level in this book or "cultural tendencies." This does not mean I disagree or take issue with Aristotle's four causes or any modern definition of the term "cause," but rather, when the idea of social causality is brought up throughout this book, it is not concerned with identifying an objective topic or specific outcomes as this idea is used to reflect cultural tendencies and belief systems. It is only through such an approach that the great historical heterogeneity inherent to human thought can be analyzed.

Contingency

The title of this book contains the word "contingency" and the inclusion of this word into the title is not random. Importantly, it does not mean the book is in anyway linked to post-structuralist thought. Post-structuralist strands of philosophy arose sometime in the 1960s as a critique of long-standing structuralist modes of thought and subsequently gained more prominence, especially in the prism of continental philosophy (which is frequently differentiated from analytical, American-styled philosophical schools in contemporary discourse). My initial interest in mechanisms and the philosophy of science stems to engagement with Glynos and Howarth's (2007) *The Logics of Critical Explanation in Social and Political Theory*, a work that can be considered to fall under the classification of post-structuralism because it built on Laclau and Mouffe's numerous works and put forward a "retroductive" (rather than inductive or deductive) form of inquiry for the social sciences. I have also published research on regime transitions, mobilization, and repression utilizing such an approach (Anisin 2014; 2017; 2021). Yet this is not the approach I adopt in this book because the topic is not necessarily a political one nor can it be analyzed through paying attention to only discourse.

Glynos and Howarth engaged with Jon Elster's widely influential mechanistic plea and deconstructed the underpinning of causal mechanisms in social science. The aim of their book was to develop an approach that could be sensitive to the self-interpretations of social actors while not reducing explanation to subjective viewpoints and qualitative cases. Glynos and Howarth's crucial contribution, alas, is rare among public post-structuralist debates and discussions, especially in academic contexts that I have come across both virtually and in person. Most post-structuralist strands of thought assume that entities cannot exist independent of discourse and that social reality (including things such as gender, race, economies, etc.) are social constructs. The different schools that can be classified under this umbrella (cultural geography; linguistics; post-marxism, feminist literatures, some psychoanalytic approaches, among others), tend to assume that all of social reality is highly contingent, and that meaning is not fixed. I agree with the latter assumption; however, I disagree with the overarching post-structuralist logic that states causality in the social world is not intelligible or identifiable. For example, a commonly held position by post-structuralist scholars is that if one assumes contingency exists (in say, the global political system), then this excludes direct causality and linear temporality (Nabers 2015). In exchange, the post-structuralist assumes that our understanding of the social world rests on difference and a lack of stable foundation(s) because of the intrinsic condition of dislocation in social structures. From these suppositions, post-structuralists attempt to investigate how discourses and identities get constructed and articulated which often results in analyses that arrive at a certain degree of validity because they capture how meanings get fixated by political actors and how they can also get disrupted and changed. Yet such approaches capture only a partial set of characteristics about the social world, and contribute little to understanding why there exist cross-generational tendencies and cross-historical stability for some discursive terms and societal inclinations.

The ontological assumptions I hold about social reality may appear to be contradictory to readers because different parts of this book emphasize contextual dynamics and the importance of paying attention to qualitative cases, time periods, and discourses, while other parts of the book attempt to abstract away into general tendencies and overarching societal dispositions. The last two millennia of history have shown that human beings have a profound ability in their usage of synthetic a-priori abilities to identify different forms of truth about the phenomenal world. Indeed, some discoveries have been overridden and often, new discoveries have proven old ones false, but the predominant pattern to scientific history is that there has been a technical, biological, and physical evolution in our abilities to observe, predict, and control the natural and social worlds (the phenomenal realm). Throughout these evolutionary processes, different forms of truth have been identified and the lack of stability of social foundations did not prevent these discoveries from being made nor did subsequent changes in social foundations and discourse nullify these truths. For example, Mendeleev's identification of the periodic table of chemical elements represents the discovery of a form of truth about the natural world. Yet in the near two centuries since this discovery, scientists have been able to go much deeper in their observations of specific elements to the point in which they now are able to detect new particles and characteristics that subsist within a given element due to advances in technological measurements (e.g. the Higgs boson particle). This does not mean that Mendeleev's discoveries are false, but rather, that there are different degrees or forms of truth in the phenomenal world.

Of my time analyzing different philosophies and writings of many thinkers, I tend to concur with Berdyaev's supposition that there are different levels and types of truth, and that truth is not and can never be a universal concept. For example, with reference to the splitting of the atom and the advent of the atomic bomb, Berdyaev noted that, "Science if it does not reveal the Truth, at least reveals truths, and our modern world is plunged ever deeper into shadow" (Berdyaev 1952, 21).

This is also why the ideas put forward in this book are not akin to the critical realist traditions which assume there can be a "meta-theory" for explaining our world (especially as applied in the context of social science). Critical realists hold ontological assumptions of akin to realism in that entities and tendencies in the world are real, but are not constitutive of law-like and functional characteristics (Bhaskar 1986; 2008; Collier 1994). An epistemological aim of critical realism is to analyze empirical reality and uncover causal mechanisms which sets it apart from the aforementioned discursive approach that seeks to "deconstruct" the meanings associated with social structures and identities. While I sympathize with the critical realist claim that not all entities are discursive constructs and that structures are ever-present in the world and get reproduced by human agency throughout time, I completely disagree with the assertion that topics such as morality (Ash 2022) or any aspect of the noumenal world can (or should) be rationally addressed.

Order of Chapters

The order and makeup of this book is as follows. The second chapter sets readers on a path towards understanding the history of social causality. In order to accomplish this, it begins with an exploration of the advent of discoveries that were made in early history and in early civilizations including Ancient Egypt and Mesopotamia. The Ancient Greeks are then given attention which includes the works of Aristotle, Plato, Pythagoras, among others. The chapter then delves into discoveries that were made during the so called "dark ages" of history followed by the Middle Ages. Afterwards, the advent of modern science is explained along with the key early debates that were held by philosophers in the 17th and 18th centuries, including Hume and Kant. The chapter finishes with a section on the emergence of social science and introduces readers to the major differences between natural and social sciences. The third chapter identifies the historical origin of the mechanism through tracing this term's etymology to the Ancient Greek context. A categorization of theatrical structure in which mechanisms were used as physical plot-shifting tools is presented through observation of early Greek theater settings. This assessment is complimented by an explanation of religious and socio-cultural tendencies of the time. The chapter reveals how mechanisms were used in some of history's most noteworthy Ancient Greek theater stories. The fourth chapter identifies the changing etymological characteristics of mechanisms in the Middle Ages in a time period that was marked by war, famine, and the eventual emergence of the enlightenment. The fifth chapter then delves into the emergence of the scientific era and the natural sciences. This is an era that marked the total elimination of theistic and metaphysical interpretations of causality as observed through the etymological characteristics of mechanisms. The sixth chapter identifies new historical processes in the specific realm of artificial intelligence (AI), and presents evidence that causality is bound to become, for the first time, independent of human cognition. Likewise, this chapter explains the advent of computation, the emergence of the digital era, and the basis of artificial intelligence. It concurrently draws on Martin Heidegger's essays on technology and identifies how mechanisms of artificial intelligence are bringing about social outcomes and causal processes. The final chapter, chapter seven, concludes and overviews the findings of this book. It overviews the implications of the book for the humanities and social sciences, and puts forward possible routes for future interdisciplinary research on social causation.

References

Anisin, Alexei. 2021. "Identity, Repression, and the Collapse of Apartheid." *The Strategic Review for Southern Africa* 43 (2): 308-336.

Anisin, Alexei. 2017. "Antagonisms and the Discursive Sedimentation of American Gun Culture: A New Framework." *Cultural Studies↔ Critical Methodologies* 17 (2): 133-139.

Anisin, Alexei. 2014. "The Russian Bloody Sunday Massacre of 1905: a Discursive Account of Nonviolent Transformation." *Politics, Groups, and Identities* 2 (4): 643-660.

Aristotle. 1966. *Metaphysics*. Ann Arbor: University of Michigan Press.

Ash, Steve. 2022. *Explaining Morality: Critical Realism and Moral Questions*. London: Routledge.

Ashby, Clifford. 1998. *Classical Greek theatre: New Views of an Old Subject*. Iowa City: University of Iowa Press.

Ballard, Paul, and Stephen R. Holmes. 2006. *The Bible in Pastoral Practice: Readings in the Place and Function of Scripture in the Church*. Grand Rapids: Wm. B. Eerdmans Publishing.

Berthold-Bond, Daniel. Hegel's Grand Synthesis: A Study of Being, Thought, and History. Albany: State University of New York Press.

Bhaskar, Roy. 1986. *Scientific Realism and Human Emancipation*. London: Verso.

Bhaskar, Roy. 2008. *A Realist Theory of Science (Second Edition)*. London: Verso.

Bell, James. 2013. *The Spiritual World of the Hobbit*. Bloomington: Bethany House Publishers.

Berdyaev, Nikolai. 1937. *The Destiny of Man*. London: Geoffrey Bles.

Berdyaev, Nikolai. 1952. *The Realm of Spirit and the Realm of Caesar*. London: Donald A. Lowrie.

Blank, Andreas. 1999. Why Do New Meanings Occur? A Cognitive Typology of the Motivations for Lexical Semantic Change, in *Historical Semantics and Cognition*, edited by Blank, Andreas, Koch, Peter, 61-90. New York: Mouton de Gruyter.

Bostrom, Nick. 2014. *Superintelligence: Paths, Dangers, Strategies*. Oxford: Oxford University Press.

Brockett, Oscar and Franklin Hildy. 2007. *History of the Theatre, Foundation Edition*. Boston: Allyn and Bacon.

Calian, Carnegie Samuel. 1969. *Berdyaev's Philosophy of Hope: A Contribution to Marxist-Christian Dialogue*. Augsburg Publishing House.

Collier, Andrew. 1994. *Critical Realism an Introduction to Roy Bhaskar's Philosophy*. London: Verso.

Collier, Andrew. 2004. *In Defence of Objectivity*. London: Routledge.

Cunningham, Maurice. 1954. Medea ΑΠΟ ΜΗΧΑΝΗΣ. *Classical Philology* 49 (3): 151–160.

Elster, Jon. 1989. *Nuts and Bolts for the Social Sciences*. Cambridge: Cambridge University Press.

Elster, Jon. 1998. A Plea for Mechanisms. *In Social Mechanisms: An Analytical Approach to Social Theory*, edited by Peter Hedstriim and Richard Swedberg, 45-73. Cambridge: Cambridge University Press.

Falcon, Andrea. 2022. "Aristotle on Causality." *The Stanford Encyclopedia of Philosophy*. Last Modified Spring 2022, Accessed March 3, 2022. https://plato.stanford.edu/archives/spr2022/entries/aristotle-causality/>.

Gerring, John. 2001. *Social Science Methodology: A Criterial Framework*. Cambridge: Cambridge University Press.

Glennan, Stuart, and Phyllis Illari. 2018. *The Routledge Handbook of Mechanisms and Mechanical Philosophy*. London: Routledge.

Glynos, Jason, and David Howarth. 2007. *The Logics of Critical Explanation in Social and Political Theory*. London: Routledge.

Grzega, Joachim. 2004. *Bezeichnungswandel: Wie, Warum, Wozu? Ein Beitrag zur englischen und allgemeinen Onomasiologie*. Heidelberg: Winter.

Hegel, G.W.F. 2018. *Hegel: The Phenomenology of Spirit*. Oxford: Oxford University Press.

Kant, Immanuel. 1797. *Critique of Pure Reason. Trans. Norman Kemp Smith, 1997*. New York: Palgrave Macmillan.

Kaufmann, Walter. 2002. *Goethe, Kant, and Hegel: Discovering the Mind*. New Brunswick: Transaction Publishers.

Kincaid, Harold. 2009, "Causation in the Social Sciences", in *The Oxford Handbook of Causation*, edited by Beebee, Helen, Christopher Hitchcock, and Peter Menzies, 726-743. Oxford: Oxford University Press.

Leuridan, Bert. 2010. "Can Mechanisms Really Replace Laws of Nature?" *Philosophy of Science* 77 (3): 317–40.

Little, Daniel. 2012. "Social Mechanisms and Scientific Realism: Discussion of Mechanistic Explanation in Social Contexts, by Johannes Persson." *Social Epistemology Review and Reply Collective* 1 (3): 1-5.

Mahoney, James. 2001. "Beyond correlational analysis: Recent innovations in theory and method." *Sociological Forum* 16 (3): 575-93.

Nabers, Dirk. 2015. *A Poststructuralist Discourse Theory of Global Politics*. New York: Springer.

Norkus, Zenonas. 2005. "Mechanisms as Miracle Makers? The Rise and Inconsistencies of the "Mechanistic Approach" in Social Science and History." *History and Theory* 44 (3): 348-372.

Obolevitch, Teresa. 2019. *Faith and Science in Russian Religious Thought*. New York: Oxford University Press.

Papadogiannis, Argyris S., Marilena C. Tsakoumaki, and Thomas G. Chondros. 2010. "Deus-Ex-Machina Mechanism Reconstruction in the Theater of Phlius, Corinthia." *Journal of Mechanical Design* 132 (1): 011001.

Popa, Tiberiu. 2018. "Mechanisms: Ancient Sources." In, Glennan, Stuart, and Phyllis Illari, *The Routledge Handbook of Mechanisms and Mechanical Philosophy*. London: Routledge.

Ratzinger, Joseph Cardinal. 2016. *Joseph Ratzinger Collected Works: Theology of the Liturgy*. San Francisco: Ignatius Press.

Roux, Sophie. 2018. "From the Mechanical Philosophy to Early Modern Mechanisms." In, Glennan, Stuart, and Phyllis Illari, *The Routledge Handbook of Mechanisms and Mechanical Philosophy*. London: Routledge.

Solomon, Robert. C. 1983. *In the Spirit of Hegel: A Study of G.W.F. Hegel's Phenomenology of Spirit*. New York: Oxford University Press.

Storey, Ian C. and Ariene Allan. 2014. *A Guide to Ancient Greek Drama. Vol. 4.* Hoboken: John Wiley & Sons.

Traugott, Elizabeth Closs, and Richard B. Dasher. 2001. *Regularity in Semantic Change. Vol. 97*. Cambridge: Cambridge University Press.

Williamson, Jon. 2011. Mechanistic Theories of Causality Part I. *Philosophy Compass* 6 (6): 421-432.

Zizek, Slavoj. 2012. *Less Than Nothing: Hegel and the Shadow of Dialectical Materialism*. London: Verso.

CHAPTER TWO

THE HISTORICAL ORIGINS OF PHILOSOPHY AND SCIENCE

Any inquiry into how social causality, and in turn, the mechanism, have historically been conceived of must begin with a time period that predates our modern period which is dominated by science, the scientific method, and experimental methodologies. If one is to attempt to grasp the outstanding differences between how causation (defined as the act or process of causing or the act or agency which produces an effect) operates in the natural and social worlds, one must not only begin at the origin of natural scientific practice, but they must also consider noteworthy developments that arose throughout a grandiose trajectory of human history which begins with the antiquity of inventions and ideas. The aim of this chapter is to unpack this heterogeneity and identify patterns that arose in different stages of human development–beginning with eras that predate the arrival of civilization and organized religion. This will enable readers to get acquainted with the cross-historical character of this book, and to set the stage for subsequent chapters in which specific eras of human cognition on causality and mechanisms will be categorized and given qualitative attention.

Throughout what truly is an immensely rich and assorted history of scientific practice, philosophers, scholars, industrialists, and enthusiasts have sought to understand the character of causation. Up until the late 19th century however, science–as we comprehend the meaning of the term today–was limited to the study of the natural world. Natural science was dominant as physical and material phenomena formed the primary basis of investigation–of chemicals (chemistry), plants, animals, humans (biology), environmental topics, agriculture, among other topics that were observed and studied outside the realm of society and social structures. Items were categorized, weighed, measured, and the empirical tools used to achieve mensuration were assumed to not be contingent on researchers' own interpretative facets or social constructions of what constitutes verifiable and measurable criteria. In the grand scheme of human history however, the systematic development of coherent bases of scientific knowledge was long coming–for tens of thousands of years prior, attempts at controlling the natural world by humans often fell short, and when looking back at human history, one would be hard pressed to state that the development of our scientific mastery over nature was a linear process.

Of all animals and subjects found in the animal kingdom, only one category–human beings, rely on and utilize fire. For every single mammal, insect, reptile, or amphibian, fire represents a grave danger, but for human beings, fire has historically functioned as a necessary condition that enabled us to exercise a degree of control over nature. For example, even apex predators (such as wolves), are scared of fire to such an extent that they will not even come near a campsite that has smoke coming from it and will go to great lengths to avoid areas that smell of smoke. Although not perceived by many as a scientific endeavor or invention, the discovery of fire led to among the greatest transformative processes in the history of the world. Fire fundamentally transformed the relation between human beings and nature. Goudsblom (1992) shows how this led to status of early human beings shifting from being subject to predators to turning into predators themselves. Fire also enabled human beings to cook and use smoke as a means of communication. The former likely contributed to greater lifespans as the problem of dangerous bacteria in food was, to a degree, combated. Correspondingly, fire enabled humans to create ceramic-like objects and engage in smelting which improved their capacity to store goods (including food), and it led to important improvements in weaponry which up until that point, was based on stone crafting.

It is estimated that some two million years ago, stone tools were invented for the first time. This invention shifted dietary patterns to meat eating–stone tools enabled carcasses of dead animals to be accessed including their internal organs and muscles. This had major consequences for subsequent development of brain size (Watson 2005, 23). Several new species (H. Habilis and H. erectus) later appeared which was followed by the eventual rise of Homo sapiens and of Neanderthals. Both of the latter species' brain sizes were markedly larger than predecessors. Monumental shifts then started to arise which can be observed throughout the following time frames. These time frames are still under debate in their respective disciplines but nevertheless constitute the best observational estimations to date: the earliest identification of cave art has been dated to some 40,000 years ago. As explained by Watson (2005), "the advent of art is so sudden (in paleontological terms), and so widespread, that many scientists think it must reflect an important change in the development of early man's mental state" (Watson 2005, 33). Art was eventually followed by the development of language (10,000 to 20,000 years ago), which was then followed by the domestication of animals (14,000 to 6,000 years ago) (Watson 2005, 53), metallurgic mastery (4000 years ago) (Watson 2005, 67), writing and scripts (5000-3500 BC), and the invention of money (3000 BC) (Watson 2005, 71).

Throughout the period of 4000-3500 BC, the first primitive religions emerged. It is likely that of the ancient burial rituals which existed tens of thousands prior, some probably did feature religious connotations. For example, if one considers that the burying of loved ones or tribal members may indicate that thought was given to what would happen to their bodies after they died, then perhaps the roots of religion can be dated to the oldest of ancient burial sites. It appears that a coherent set of religious thoughts and ideas only emerged alongside the advent of writing and scripts. In 2900 BC, the epic of Gilgamesh (in the context of Babylon) gained widespread prominence. Mesopotamia (the area of modern Iraq) produced the first libraries and collections of scripts. In these eras, attempts at explaining natural phenomena were deeply mystical in their characteristics (Ledoux 2002, 34). Different ancient civilizations (such as Ancient Chinese, Egyptian, or Indian) carried out practices that should be considered among the first scientific practices in human history. These included anatomical surgeries; gauging blood flow and heart function; examining natural events such as storms or earthquakes, among numerous other activities that were truly captivating for that era in time. For example, the Ancient Hindus engaged in an array of different operations that can considered to be scientific by today's standards including metallurgical purification of metals; tracing planetary motions; and carrying out cataract surgeries (Kumar 2014). Up until the era of Ancient Greece and the specific contributions put forward by Aristotle however, there arguably was no coherent enterprise of scientific practice. Most attempts at rationalizing natural phenomena were rooted in magic. The Ancient Egyptians and Mesopotamians took on the following mystical and supernatural approaches, "Egyptians and Mesopotamians viewed the world as a place where magic was essential for survival. It was used to explain virtually all phenomena that we would regard as natural. This is nowhere better exemplified than in medicine, where the diagnosis and treatment of internal ailments relied heavily on magic" (Grant 2007, 3).

This all changed with the rise of natural philosophy. Natural philosophy predates natural science and can be thought of as a precursor to it. Some have attributed great historical importance to natural philosophy and have described it as being the "great mother of the sciences" because by the 19th century, it led to a systematic development of physics, chemistry, and biological sciences (Grant 2007). Natural philosophy, as described by Grant (2007), encompassed all natural phenomena found in the physical world and it sought to discover physical causes of natural effects. In its context of origin however, natural philosophy was not reliant on mathematics and at the time, it functioned and was implemented outside the realm of astronomy, optics, and mechanics–largely due to its concern with physical causes (Grant 2007). The Ancient Greeks articulated a plethora of revolutionary ideas about human beings' position within and in relation to nature. For the first time in human history, a systematic effort was put forward based upon the premise that our cognitive abilities can lead us to explain natural phenomena. The specific historical era that this began to occur was from 500-600 BC. This era turned out to be fundamentally transformative. This was also the first attempt to form a coherent framework about the phenomenal world and was the context that the mechanism was first conceived.

Ancient Greeks and the advent of Natural Philosophy

At this time, it was not the case that supernatural or mystical explanations of natural phenomena were eliminated or declined. As the subsequent chapter on mechanisms in Ancient Greece will reveal, such explanations were still hegemonic in their societal function. What did arise nevertheless, is the pre-Socratic period in which philosophers (e.g. such as Thales of Miletus) began to make among the first distinctions between the natural and supernatural (Lloyd 1970). Thales of Miletus put forward the first logical account of the universe and was arguably the first person to have rejected superstition by arguing that the world could be accounted for and explained by reason (Guthrie 2012). Although Thales' (and many of the subsequent figures') explanations for what constitutes the universe turned out to be incorrect if assessed from our current modern body of materialist knowledge and experimental science, the conceptualization that went into developing their arguments was of great significance for fruition of scientific thought. Along these lines, the pre-Socratics ended up shaping the discipline that would eventually be known as natural philosophy (Grant 2007).

Subsequent contributions and ideas by the Pythagoreans (who some believe introduced the first mathematical theory), by Parmenides, Zeno, Leucippus, Empedocles, Luecippus of Elea, and Democritus all provided what at the time were historically new modes of thinking about the natural world that were outside of the realm of myth and mysticism. While Parmenides was likely the first person to ask how atoms can be separated from one another, it was the latter of these figures, Democritus, who put forward the first ordered theorization of atoms by accounting for how atoms could separate and how they could move. It was demonstrated that physical movement was based on either the union or separation of unchangeable atoms (Van Melsen 2004, 19). Democritus' ideas on atoms and the void turned out to be significant because they were underpinned by the assumption that there is no theistic or divine force behind why phenomena occur in the natural world (Experts 2019, 374). Next came Pythagoras, a name that still is attached to his theorem and is taught in mathematical instruction to young people in schools across the world. As described by Watson (2005), due to their fascination with numerology, Pythagoreans were the first to identify that the earth was a sphere (Watson 2005, 130).

Plato and Aristotle

By the 4th century BC, Plato and Socrates appeared and at the time of writing this book, these figures remain among the most widely discussed philosophers of our time. As a pupil of Socrates, Plato's ideas and contributions ended up seeing greater interpretation in political discourse (and politics more generally) than in the natural sciences. This probably stems to the fact that Plato emphasized the concepts of art (techneˆ) and beauty. His contributions were more mystical than rationalist in character. In contrast to natural forces, art and beauty were assumed by Plato to be superior outcomes of human experience and constituted our highest cognitive potential. The purpose of art for Plato was to apply objective principles to structuring cosmic nature. Art enabled human beings to attempt to obtain cognitive insights into cosmic nature (which Plato assumed was ruled by a theistic power). Art was expected to be able to provide human beings with an opportunity to deliberately attempt to acquire self-mastery, which would in turn, enable us to control our own patterns of behavior that were not limited to a given brash form of individualism, but rather, constituted self-government, statecraft, and "biosocial necessities of humanity" (Lodge 2014, 45). Plato's emphasis on *form* as constituting the ultimate component(s) that structures human knowledge meant that the human mind can only construct an ideal form of perceived phenomena of the physical and material world. Since forms are not contingent or reliant on sensory knowledge, they can thus only be apprehended by one's intellect. Plato's forms seem to be timeless. They are eternal and everlasting universals that are non-definable in observational terms (Kouremenos 2018).

Analytically, Plato drew from Pythagoras and believed the nature of the universe is triangular. This led to what later were referred to as platonic solids which were comprised of five geometric shapes that are identical to one another (in meeting at the same dimensional angles). These include the cube, the octahedron, the tetrahedron (or pyramid), the dodecahedron, and the icosahedron (Sutton 2002). Subsequent to Plato, the temporal order of any historical overview of Ancient Greek thinkers would lead to Aristotle. Aristotle was a student of Plato while in Athens and over the course of his life he ended up producing more than 200 different significant works on a vastly heterogeneous collection of topics ranging from biology to ethics, formal logic, geometry, meteorology, animal behavior, anatomy, and law and politics. Many of these works were not dated and hence, still to this day, debates ensue with regard to their chronological order. Likewise, of the great body of work that Aristotle produced throughout his life, a lot of written material was lost, while most of his lectures were preserved.

Aristotle was arguably the first to develop the concept of matter, which he believed was vital for understanding the natural world (Ebry 2015, 3). Marie Leroi (2015) in *The Lagoon: How Aristotle Invented Science*, puts forward a compelling account of Aristotle's importance to the development of science and emphasizes the vast number of questions Aristotle not only posed, but answered. Aristotle accurately described and detailed the anatomy of over 100 different species of animals. Alongside this fascinating biological itinerary of research, arguably his two greatest contributions were the 1) conception of the syllogism. At the time, this articulation was revolutionary because it presented the first deductive framework of analysis. This brings us to: 2) posterior analytics, which set out a scientific methodology based upon mathematics. In addition to these two articulations, Grant (2007) describes Aristotle's contributions as follows,

> "Aristotle constructed a system of the cosmos that endured for more than two thousand years in three different civilizations and cultures. His discussions of motion in the Physics set the stage for subsequent controversies that resulted in medieval advances and ultimately led to Galileo and Newton" (Grant 2007, 36).

In multiple respects, Aristotle ended up reversing Plato's thoughts by 180 degrees. Instead of forms and mystically oriented frameworks, Aristotle put forward the first deductive approach to the study of phenomena and while doing so, he categorized everything and anything he believed was worth studying. Along with the aforementioned (chapter one) Four Causes, Aristotle also put forward what turned out to be the first conceptualization of different forms or types of science – including the practical sciences (associated with principles that are put forward by doers in a given field), the productive sciences (associated with principles that are put forward by producers in a given field), and the theoretical sciences (associated with principles put forward by observers in a given field) (Johnson 2015, 174). Johnson (2015) overviews the utility of this trichotomy for discussing the objects of sciences and categorizes them as follows, (drawn from Johnson 2015, 174):

(1) Theoretical Sciences:
   a. Theology (metaphysics, the science of first principles and causes)
   b. Mathematics
   c. Natural Science
(2) Practical Sciences
(3) Productive Sciences

While disputes have arisen about the importance or order that Aristotle attributed to each of his three scientific categories (e.g., Johnson engages in several debates on this topic and argues that productive science is subordinate to practical science which is itself subordinate to theoretical science), we must simply step back and consider the vastly influential implications that arose from these categories. Through putting forward these categories, Aristotle denied one could arrive at or produce scientific knowledge by means of sense perception alone (Grant 2007, 39), which entails that universal propositions are essential for the production of scientific knowledge. As readers will observe later in this book, Kant's distinction between the noumenal and phenomenal worlds added an intriguing twist to Aristotle's framework of categories of scientific practice. In *Physics*, Aristotle argued that nature is a cause that operates for a purpose which he defined as "a principle of motion and change" (Grant 2007, 410). This "all natural motion" was said to order the movement of stars, the heavens, and corporeal elements. Combined with his framework of the Four Causes, Aristotle set the stage for the next two millennia of scientific development and practice.

Discoveries and Ideas in the Dark Ages

After Aristotle's death, much of his work was preserved by Plotinus and later, the Neoplatonists. This school of thought reemphasized Plato's forms while simultaneously attempting to reconcile Plato and Aristotle's work with one another. Over the course of the next 500 years, the ideas of Aristotle, Plato, and all of the Ancient Greek philosophical hegemons were disseminated and translated into varying areas of the Middle East. Here, the rise of Nestorianism and Monophysite translators resulted in new discursive constructions that made their way into Persia, the Byzantine Empire, Baghdad, and into Ancient Syria. In these contexts, there was hostility toward natural philosophy that came out of Greece. Grant (2007) argues that such hostility was common and viewed with suspicion because of what many believed to be natural philosophy's link with subversive religious themes–themes that were at odds to Islamic higher education practices. Along similar lines, in the Chinese context, hugely innovative tendencies can be observed across the entirety of the Song dynasty (960-1279). The degree of technological advancement that arose in this period of Chinese history far surpassed any other region of the world as the Chinese had already developed paper; printing; iron and steel production and even coal mining. Steel was already being produced during the Song dynasty (Watson 2005, 304). On top of these technological advancements, the Chinese were the first to discover and use gunpowder and develop methods of sailing that at their time of creation, were unmatched in terms of the quality of their maritime expeditions.

Only into the 12th century did Aristotle's and other Ancient Greek texts begin to enter into Europe through Latin translations. In Oxford, Robert Grosseteste not only produced the first translations of a number of different important works, but many consider him to be the inventor of the experimental method (Watson 2005, 329). Grosseteste specifically developed a model of induction and improved on knowledge of how to measure phenomena which Watson (2005) contends led to "profound psychological and social change" in the 13th and 14th centuries. During this time period, the clock was invented along with other measurement tools (Watson 2005, 329). We must consider a commonly unappreciated fact pertaining to what many have incorrectly referred to as the "dark" or Middle Ages (all the way into the 12th century). Here, western civilization was not as stagnant as usually assumed with regard to discoveries and scientific and philosophical activity. Before identifying the advent of the clock and the tracking of a daily work hour in Europe, important developments were taking place across in contexts throughout the Middle

East. In the 8th century, a Persian mathematician named Muhammad ibn Musa al-Khwarizmi created the first comprehensive works on linear algebra and quadratic equations. Al-Khwarizmi's contributions resulted in the transformation of a concept of a number from an arithmetic character (as a fixed quantity) into a variable in an equation (Al-Daffa 2020). As described by Al-Daffa (2020), Muslim mathematicians in the 8th and 9th centuries laid the foundation for solving linear and quadratic systems (including third- and fourth-degree equations) by systematizing the science of algebra and preserving its links to geometry. Muslim scholars additionally made substantial contributions to trigonometry and geometry.

In Europe during this time, greater productivity started to arise in agricultural spheres and more people started slowly moving into urban areas. The invention of the water mill (estimates indicate the 6th century) was of great significance as it led to greater productivity in the manufacturing of wool and cloth products. Along with water mills, the plough was made, crop rotation systems emerged, along with horse shoes and neck harnesses (for horses)–all of which increased productivity (Watson 2005, 326). In the latter half of the 13th century, the clock was invented in Europe and working hours began to be timed by the sound of bells that would ring in village sand town centers (Watson 2005, 329). On a personal basis, by this time in European history, an extraordinary development arose–people began viewing themselves as individuals (in Western parts of the European continent) due to a transformation in how Christianity was interpreted. Already in the 13th century, certain universities such as Oxford were host to scientific movements. From 1300-1500, some of the most historically salient and well-known institutions were founded in Western Europe which gave rise to new autonomy for scientific and philosophical organizations.

The manner of how numbers (and analytical reasoning more broadly) were conceived of and used in daily life became more specific through improvements in measurement capabilities. This resulted in greater particularity in all spheres of social organization and what Watson (2005) describes as the growth of quantification in record keeping (Watson 2005, 381). The cherry that ended up being placed on the top of this cake of historical innovation was the advent of printing. Manufacturers arose that produced accessible paper which in turn led to printed books, instructional guides, and more information being made available to publics than ever before. For this first time in history, wide spread access to non-collective forms of knowledge and information arose. Then came sea-based exploration which became rampant across what today are known as Italy and Spain. By the early 1500s, Western European explorers (such as C. Columbus) were financed by monarchies to sail off to faraway lands and identify new opportunities for resource extraction. These processes led to massive transformations in sailing technology, armament, geographic tracking, among many other technical improvements to human behavior including the field of medicine (particularly in anatomy).

A nail went into the coffin to the Middle Ages when the Italian renaissance arose. Here, great amounts of wealth were accumulated in different provinces of the Italian portion of the remnants of the Roman empire. In conjunction with the vast amount of different technological advancements that were made from the 13-15th centuries, many Italian cities were largely population (upwards of 50,000 to even 90,000 inhabitants), had universities, and were connected to far-away lands through trading routes. In their own independent processes of knowledge formation, philosophers of the Italian Renaissance such as Vlala, Petrarch, Ficino, or Telesio contributed what would turn out to be the bedrock of humanism (Kristeller 1964). The emergence of Renaissance humanism bestowed a new hegemonic overtone that carried over into the age of industrialization.

The Mechanization of Nature and the Emergence of Science

The age of medieval Europe was not totally stagnant and infertile when it comes to scientific discoveries. There were advances that were made in mechanics, optics, and mathematics. One must also attribute importance to the influence of Christian interpretations of Greek scientific discoveries and natural philosophy as a whole. The early Church Fathers (such as Tatian, Theodoret, or Saint Basil) viewed Greek science as pagan and attributed it to disinformation and confusion (Grant 2007, 239). At the end of the day however, the hostility towards Greek science and philosophy did not triumph. In Latin contexts, the influence of Thomas Aquinas, St. Augustine, and Albertus Magnus were monumentally epochal because they put forward variant rationalistic viewpoints on natural philosophy and related it to faith and theism. Although it would be difficult to call their contributions a reconciliation of Christianity and science, as one would expect, much of what they discussed stemmed to Aristotle, especially his work in *Physics*. Magnus' ideas were especially transformative because they put forward a conceptual distinction between knowledge that was theistic and knowledge that was natural. As described by Watson (2005),

"Albertus set loose a change in the world, the power of which he couldn't have begun to imagine" (Watson 2005, 330).

Once historically transformative discoveries were made by Copernicus, Galileo, Kepler Newton, and numerous others, Aristotle's ideas on the cosmos were overturned as was his framework of syllogistic logic. Alongside what many have categorized as a scientific revolution, there also was a philosophical revolution–the emergence of liberal philosophy fundamentally transformed prior philosophical thought. Figures including Thomas Hobbes, John Locke, Francis Bacon (and many others), put forward new ideas about how empirical reality can be studied in light of newfound knowledge of the natural world which could be measured with an abundance of newly created instruments. Of equal importance was the Protestant reformation which changed the institutional basis of Christianity across much of Central and Western Europe. The rise of humanism also complimented the general tendency of a move towards human autonomy and conceptions of individualism which hitherto were not ever developed or conceived of.

One helpful approach to understanding these historical changes in human beings' relation to the natural world can be observed in a framework that was put forward around 75 years ago by philosopher Nikolai Berdyaev. Berdyaev categorized meta-historical processes of change with reference to human beings' conception and relation to the natural world, to religion and to technological change. At the time of writing, Berdyaev (1952) argued that there five periods in all of history (the fifth of the four was to take place in the future). Period one was marked by human beings' submersion to cosmic life–it was a time in which human beings did not yet conquer nature and were living in highly primitive forms of social organization. In this period, human beings were confined to beliefs that were pagan in that their relation to theistic deities was contingent on linkages of gods to different forces and elements of nature. In the second period, human beings became freed from cosmic forces and this period importantly saw the onset of elementary forms of economics including serfdom. The second period was importantly marked by the emergence of Jesus Christ and subsequent formation of Christianity. Further, in the third period, human beings conquered nature through mechanization and achieved scientific and technical control. This period was marked by the development of industry, capitalism, and the solidification of science. Put differently, this is the period in which a fully coagulated prototype of science (as it is conceived of today) was created. Next, the fourth period (our current era) has been marked by what Berdyaev described as the disruption of cosmic order, the splitting of the atom, advent of nuclear weapons and technology, and the dissolution of organic forms of social organization. It is also the era in which autonomous forms of technology have appeared – Berdyaev predicted the rise of artificial intelligence long before it actually was created (Anisin 2021), which resulted in the subjugation and enslavement of human beings to our own discoveries. Lastly, the fifth period will take place sometime in the future and in this period, Berdyaev predicts that an eschatological revolution will arise featuring a spiritual transfiguration.

Later chapters of this book will turn back to Berdyaev's helpful categorizations. For our immediate purposes, we must consider the cogency of particular claims laid forward by Berdyaev that characterize the third period. Historians and philosophers share a near-universal recognition of the idea of a scientific revolution–sometime between during the 18th century, massively transformative changes began to arise that only intensified as the next two centuries went on. By the 1859 publication of Darwin's *On the Origins of Species by Natural Selection*, science was solidified and the old religious, traditionalist viewpoints of the natural world were nearly done away with in upper societal ranks. Henry (2002) makes a plausible argument in that the emergence of the scientific revolution in the 17th and 18th centuries ended up fostering the basis of modern society as we conceive it today. The scientific revolution brought about changes in human thought, behavior, and societal organization that were totally new in their lineament. This revolution constituted a "very real process of fundamental change" (Henry 2002, 2), and a process that not only was symbolized by a radical break with the past, but one that advanced human beings' understanding of nature and natural change. Henry (2002) describes this transformation as follows,

> "A simple but essentially accurate way of summing up what took place in the Scientific Revolution, then, is to say that the natural philosophy of the Middle Ages, which had tended to remain aloof from mathematical and more pragmatic or experiential arts and sciences, became amalgamated with these other approaches to the analysis of nature, to give rise to something much closer to our notion of science" (Henry 2002, 5).

Henry goes on to argue that it was not the case that natural philosophy simply morphed into what we understand or refer to in our conceptualization of the emergence of science, but rather, natural philosophy was a term that was used to describe understandings of the physical world. By the beginning of the 19th century however, the term science obtained a specific set of meanings that we continue to attribute to it today.

The Characteristics of Natural Science

Concurrent to the revolutionary basis of the advent of modern science was the idea that the natural world could be studied mathematically–this idea was solidified by the end of the 17th century as the discoveries of Galileo, Newton, and Kepler shifted and displaced previous understandings of the world (Shapin 2018). The period of the renaissance also led to significant changes in societal preferences across much of Europe wherein increased demands emerged for practical innovations. Henry (2002) associates this increase with a break of tradition and the onset of cultural relativism. More and more people began to lend their support to new ways of thinking and new ideas. Perhaps the greatest symbol that emerged out of this new line of thinking can be observed in Immanuel Kant's contributions to philosophy and science. Kant critiqued rationalist and empiricist frameworks that had arisen from the enlightenment and thereafter. Kant's ideas transformed how human beings perceived themselves and their mind. He articulated a new type of empirical realism by arguing against previous ideas put forward by great minds such as Locke, Hume, Berkeley, Spinoza, Leibniz, and Descartes. Likewise, many believe Kant reconciled both empiricism and rationalism, which during his lifetime, were at odds with one another. Furthermore, Kant's contributions have often been situated within the larger historical and contextual era of his life in German Idealism. Along with Kant, idealists assumed that our perceptions of the world are contingent on the manners in which they appear to us and our perceptions.

Kant argued that one cannot possess knowledge of a given empirical phenomena that is independent of the conditions that are tied to our experience of that phenomena. In other words, the knowledge we have of phenomena are limited to the conditions of experience that we attach to an object or process. Among the most cited aspects of Kant's thought is in his categorization of the noumenal and the phenomenal realms. The latter contains empirical phenomena that we can perceive, while the former constitutes a higher set of entities that are beyond human cognition. In his framework laid out in the *Critique of Pure Reason* (published in 1781), Kant argued against Hume's views of causality in which it was contended that causality is incapable of being proven due to our propensity of misinterpreting it. For Hume, the principle of causation determined all scientific and empirical reasoning. Hume contended that the causal principle is synthetic in character and that empiricism must ultimately lead one to skepticism (Ewing 1969). Kant partially adopted these assumptions and then provided an alternative (logically based) set of solutions that set the stage for a hugely shifting set of ideas that went onto change the structure of philosophical (and scientific) thought forever. Kant contended that the previous a-priori and posteriori truths that were popular in both rationalist and empiricist frameworks were incorrect.

Kant assimilated human judgment to synthetic and analytical processes. This broke the old view of judgment being conceived as primarily analytic in character (Ewing 1969). In Kant's two categories of judgment (the synthetic and analytic), he argued that human beings can articulate knowledge that is true from an a-priori perspective (which is particularly complementary to our ability to engage in mathematics). For instance, Kant argued that constructing ideas about the angles in a triangle can be done in exactly the same way regardless of the type or age of the person that engages in this activity. This means that the a-priori category of knowledge is, in a sense, universal because it is not contingent on experience. This is particularly why Kant was inclined to postulate that a transcendental foundation must be in place for science and that such a foundation must be based on mathematics and our synthetic a-priori abilities which humans possess without necessary reference to empirical experience. The implications arising out of Kant's conceptualizations meant that major scientific discoveries were brought on by the construction of concepts by individuals who exhibited their a-priori intuitions (whether analytic or synthetic types) which then correctly corresponded to the concepts under attention (Cohen 1985, 245).

Ewing (1969) offers a helpful explanation of Kant's framework, "consciousness of the forms of perception, space and time, even apart from any particular perception of objects, implies the synthesis and the transcendental unity of apperception. Synthesis is viewed as a condition by which alone perception itself is rendered possible" (Ewing 1969, 49).

Along these lines, it is plausible to believe that the two most salient factors that human beings could determine and construct a-priori were what Kant referred to as space and time. Both space and time are understood by the human mind without necessary experience and function as "pure manifolds" because they are involved in all perception. Furthermore, Kant assumed that science progresses through massive changes brought on by intellectual revolution. The first revolution, argued Kant, was in the development of mathematics and the specific ability to measure the earth through deductive frameworks (Cohen 1985, 244). The importance here is not in the commonly ridiculed point attributed to Kant (that he conceived of his own thought as revolutionary), but in that Kant demonstrated that human cognition determines representations of the world. This meant that along with space and time, concepts such as multiplicity, unity, possibility, and causality were also contingent on our own a-priori perceptions.

Kant's views on causality must be emphasized because he described that our own synthetic a-priori conception of a cause necessarily brings about a connection with an effect. More broadly, Kant maintained that all cognition implies a synthesis (Ewing 1969, 48). As follows, Kant demonstrated that we can never expect to build knowledge of a given object outside of or external to the relation that it has to all subjects. Kant's transcendental deductive framework implied that all of our judgments must either involve a reference to the physical world or to the empirical self or to both (Ewing 1969, 55). His ideas spurred new ways of thinking, posed new questions, and eventually fostered entire fields of study that are still relevant today. In a sense, Kant opened up the black box of human cognition by drawing from theistic, atheistic, natural philosophical, and metaphysical perspectives at a time when among the fastest and most profound changes to human thought and social organization were arising.

Subsequent to Kant, a colossal number of new ideas and scientific advances followed throughout Western Europe. Abraham Gottlob Werner identified tens of thousands of rocks and put forward among the first geological surveys, in turn, this resulted in the forging of the discipline of geology. Carl Friedrich Gauss put forward a number of simple, but immensely revolutionary calculations and axioms that advanced the status of mathematics forward. Alexander Von Humboldt journeyed across the world and produced first comprehensive survey and categorization of plants (botany), atmospheric phenomena (such as air pressure and temperature), geography, and geology. By the mid-19th century, provinces in what later became Germany started seeing increased societal enrollment in university and scientific seminars began to emerge with regularity in legal, medicinal, and theological faculties. This resulted in the modern professionalization of the scholarly profession (Watson 2005, 233). On top of these new developments and tendencies, Darwin's *Origin of Species* solidified the status of natural science. By 1890, a number of biological discoveries were made that ended up changing the basis of medicinal practice. August Kekule discovered the molecule and the advent of organic chemistry followed. Along with the molecule, scientists in the 19th century discovered began to widely study (with methodological precision), mechanics, electricity, optics, and physics including the thermodynamics (Watson 2005, 342).

Infectious diseases and viruses were also addressed in systematic ways for the first time. At the turn of the 20th century, the discoveries of Robert Koch changed the practice of medicine. Decades earlier, Koch identified numerous deadly and infectious bacteria (tuberculosis and cholera) and these discoveries ended up transforming medicine. The industrial manufacturing of medicine, machines, weapons, and consumer items ranging from automobiles to typewriters led to the rise of a highly connected and new form of human organization which was already visible before the outbreak of the first world war. The natural sciences were solidified and included the following disciplines: astronomy which is based on the study of space and the cosmos; chemistry which is based on the study of substances (analytical, organic, inorganic, physical forms of chemistry); earth sciences which are based on the study of the planet (geology, oceanography, among others); physics which are based on the study of matter and energy (classical physics tend to refer to frameworks that predate the 20th century, while modern are post-1900 and include Einstein's theory of relatively as well as quantum mechanics, string theory, among others).

The Emergence of Social Science

Today there are over a dozen different disciplines (and even more sub-fields) that are classified under the umbrella of social science which include: economics, sociology, political science, anthropology, business, communication, criminology, demography, development studies, education, environmental studies, geography, gender studies, law, library science, linguistics, psychology, public administration, social work, among others. The institutionalized study of society and social phenomena arose at a much later point in human history than many would assume. It was only in the late 19th century where comprehensive works on social order started to emerge with consistency. Indeed, sociological approaches to study society can be observed in early periods of modern history through contributions made by figures such as Giovanni Battista Vico (1668-1744) (who many refer to as the "Father of the Social Sciences" (Sharpes 2009), or August Comte (who created the term sociology and arguably coined the positivist tradition), or Marx (the Communist

Manifesto), A. Smith (Wealth of Nations), Weber (Theory of Social and Economic Organization; The Protestant Ethic and the Spirit of Capitalism) Saint-Simon (Nouveau christianisme), Spencer's (Social Statistics), or Durkheim's (The Rules of Sociological Method) work, but it was really the institutionalization of scholarly research that fostered variant social scientific disciplines including sociology, economics, political science, and eventually many others. The mid-19th century saw astounding changes in much of the Western world that arose due to industrialization and urbanization. Coinciding with these processes were nationalist tides, atheism, individualism, and the formation of new political identities. Not only did these outcomes lead many to ponder over a new dawn or era of human organization and statehood, but they also were accompanied by the expansion of scholarly research throughout universities in places such as Germany, France, Britain, and the United States (among others).

The expansion of academic institutions led to greater opportunities for ideas to emerge and more resources were funneled into ideas and inquiries on social phenomena. Likewise, the institutionalization of social research was also being carried out in places such as the US to survey citizens on a variety of different issues. Surveys enabled public policy makers to gather information about citizens' preferences in new ways. For example, Booth's survey of poverty and unemployment in London was among the first of its kind and exemplifies how a trend that already began to pick up in the US (US social survey) and UK (UK social survey) in the last decades of the 19th century (Bulmer, Bales and Kish Sklar 1991). By the 1940s, the modern social survey was forged and surveys started to have an impact on public opinion, debates, and politics. As more data became available on the socio-economic characteristics of populaces, numerous new works were put out that were more rigorous than their predecessors. One particular figure who may not receive as much attention as he should in contemporary social science disciplines is Peter Drucker. With books such as *The end of Economic Man* (1939) and *The Future of Industrial Man* (1941), Drucker put forward what at the time were sophisticated attempts at generalizing about socio-economic conditions. Another monumentally innovative work can be observed in Rudolf Carnap's *The Logical Structure of the World* (1928) which set out a position that was popular among members of the Vienna Circle. Carnap later wrote the Logical Foundation of Probability (1950), a work that was also of great importance and one which we will turn back to subsequent paragraphs.

The philosophical movement advocated by the Vienna Circle (founded in 1928), played a huge role in fostering the eventual acceptance of a mathematical approach to study social phenomena. In Kantian terms, the social aspect of the phenomenal world started being studied through human beings' synthetic a-priori knowledge. Moreover, the Vienna Circle was known for its empiricist aims and logical positivism. The latter enabled it to put forward appealing and progressive arguments against metaphysical approaches (including those found in German Idealistic frameworks). The Vienna Circle also featured Karl Popper – a philosopher's whose contribution to 20th century debates in the philosophy of both science and social science ended up determining how many classify or categorize inquiry based on whether it is scientific or unscientific (via demarcation and falsification). Although Popper did not fully agree with other notable figures of this circle including those such as Ludwig Wittgenstein, Carl Hempel, A. J. Ayer, and Rudolf Carnap, coherent tendencies about what this circle assumed cab be observed. For example, a helpful classification put forward by Gorton (2012) sheds light on the tendencies of the Vienna Circle. This group of intellectuals emphasized that the following tenets are necessary for scientific inquiry: 1) sensory data – data gathered by the senses constitutes the foundation of our knowledge of the world; 2) verificationism – statements or theories that are worthy of being called scientific must correspond to empirical facts through observation and experiment; 3) antimetaphysics – unverifiable statements are meaningless; 4) antirealism – unobservable entities are at best fiction and do not really exist; 5) skepticism about causes – necessary connections that lie between events cannot be demonstrated empirically (i.e., one event causing another is nothing more than a claim that the first event precedes the second event); 6) support for deductive-nomological (covering law type of explanation); 7) unity of scientific method – the six principles above embody the only correct or true path to knowledge about both the natural and social worlds (Gorton 2012, 25).

A number of different mathematical innovations were being made throughout the first part of the 20th century and these innovations later contributed to the formation of different fields of knowledge that became known as Game Theory, Bayesian statistics, and general advancements on probabilistic causation. New analytical discoveries started emerging in the beginning of the 20th century and really picked up by the mid-20th century including noteworthy papers and books which shifted understanding about the possibility of synthetic reasoning. These included: Zermelo (1912, perfect information); Borel (1927, mixed strategies), Von Neumann (1928, minimax theorem), Von Neumann and Morgenstern (Theory of Games and Economic Behaviorm, 1944), John Nash's work on non-cooperative games and the identification of the Nash equilibrium, Arrow's Social Choice and Individual Values (1951), Black's Theory of Committees and Elections (1958), Anthony Downs' An Economic Theory of Democracy (1957) Schelling's The Strategy of Conflict (1960), Buchanan & Tullock's The Calculus of Consent (1962), Harsanyi's (1964) work on "Nature," among many other analytically based contributions to the study of human behavior.

Eventually, this heterogeneous collection of contributions to mathematics led to a revolution in statistics in social science because the "old" form of frequentist statistics was replaced by Bayesian statistics. Bayes' theorem is arguably the greatest contributor to what makes contemporary social science possible because it provides a mathematical foundation that is used to account for probabilistic causation, and in turn, has enabled researchers to make causal inferences based on assessments of large sets of data. Todays sophisticated and rigorous empirical approaches to the study of social phenomena would not have been possible had it not been for these antecedent advances in the basis of methodology – mathematics and statistics. It is also pertinent to ideate that the aforementioned contributions were not only relevant for the subsequent development and standing of social science as a whole, but they were just as relevant for natural scientific disciplines as statistical assessment of data is not inclusive to either the study of the natural or social phenomena. While social scientists widely acknowledge that their ability to predict and generalize about social phenomena are not as strong as their counterparts in the natural sciences, there have been long stemming debates and deeply rooted cleavages when it comes to the methodological strategies that should or should not be utilized to study empirical phenomena.

Differences between Natural and Social Science

One common viewpoint is that regardless of whether a given science is categorized as being natural or social, the aim common aim of scientific practice is to identify and discover laws that "cover a wide range of specific phenomena" (Gordon 1991, 34). In the social sciences, such goals are attempted to be met by way of generalization and importantly, one must consider that there are different types and kinds of generalizations. For instance, Gordon (1991) puts forward several useful distinctions on nomological propositions. First, Gordon points out that there are purely empirical laws which are analytical in nature. Second, there are also laws that involve the concept of cause (Gordon 1991, 34). Third, there are also causal laws that are based on the expectation of bringing about a given phenomenon. The first of these laws, the empirical law, stems to John Stuart Mill's framework in which an empirical law was defined as, "an uniformity, whether of succession or of coexistence, which holds true of all instances within our limits of observation, but is not of a nature to afford any assurance that it would hold beyond those limits…" (Gordon 1991, 34). For example, the historically salient political concept of the "iron law of oligarchy" (Michels 1911) which argued that social organizations tend to feature hierarchies that arise out of natural inclination rather than socially engineering or institutional building. Yet as Gordon (1991) correctly points out, such laws are contingent because even if in one single case in which an outcome is predicted or covered by the law does not occur, then this destroys the law's claim to generality.

This brings us to a crucial distinction between the natural and social sciences – the former are stable, concrete, and materially and physically solidified while the latter are contingent and more probabilistic in their nature. In terms of the major differences between natural and social sciences, Gordon (1991) identifies six major distinctive characteristics. The first is that 1) nomological propositions are only possibly with reference to phenomena that have a reasonable degree of uniformity. In chemistry or biology, members of the same type of cell or species tend to have perfect uniformity, whereas in the study of social phenomena, there is such diversity that nomological propositions can frequently fail to be met (Gordon 1991, 51). Second, the social sciences can hardly produce controlled experiments. Due to the time of the publication of Gordon's book (1991), it does not acknowledge that there has been progress in experimental social science (Morton and Williams 2010). Nevertheless, one would be hard pressed to believe that the basis of experimental social science is not flawed. Social interactions cannot be confined to a given setting or computed in the same way experiments on natural phenomena can. Third, progress is indefinitely bound to be slower in the social sciences when compared to the natural because it of the lack of ability of a social scientist to pursue his/her curiosity. Although Gordon does not situate this claim with reference to Kant's synthetic a-priori framework, it is plausible to assume that the following statement made by Gordon, "In view of how much of the nomological content of the natural sciences has resulted from the pure curiosity of the scientist, it is not surprising that, in this respect, progress in the social sciences is slower" (Gordon 1991, 52), is underpinned by a Kantian supposition. Fourth, the social sciences are more reliant on and prone to value judgments than the natural sciences. There is greater detachment in natural science by researchers morally. Indeed, this does not mean that natural science is by any means perfect. Fifth, in social science, the study of human behavior is fundamentally different than phenomena that are studied in the natural sciences such as rocks or planets. The specific differentiating factor here that is most significant is that the former (human beings) possess consciousness, whereas the latter (planets, rocks, or even lower-level organisms) do not. The inner mental states of human beings have no counterpart in the material world (Gordon 1991, 52). Finally, the sixth difference is in the fact that whole-part relationships in the social world are fundamentally variant when compared to those in the natural world. For example, Gordon describes how the characteristics of chemical elements are not dependent on the compounds of higher-level entities of which they are parts. The human individual, in contrast, is dependent, and by some measures or theoretical frameworks, is greatly dependent on his/her social environment.

In conjunction, these six differences have to be considered by scholars and students alike because they constitute a very practical set of barriers between natural and social science – most of which will never be overcome. This is not to say, however, that sophisticated, statistical, and most recently, computational methodological bases of social science provide us with little utility or are unfitting. Rather, acknowledging differences between the natural and social worlds is necessary for conceptually sound inquiry to be carried out. This brings us to the topic of mechanistic explanations which will be briefly addressed here (an extensive analysis is found in chapter five).

Mechanistic Explanations and their Historical Variance

The probabilistic revolution in statistics led to creation of many new methodological tools and interestingly enough, in the subsequent decades (post-1960), computers started emerging as research sources in government institutions and universities. Eventually, computers became consumer items. This important inclusion of computation led to the collection of never-before-seen amounts of data and data sets, many of which became publicly available. In conjunction, the advent of Bayesian statistics along with computational power led to grandiose ideas about advancing the status of the social sciences into a higher realm–one that was thought to be closer to the natural sciences. Social scientists gradually moved away from description and into the realm of prediction and causal inference. For example, Gerring (2012) surveyed the basis of argumentation in articles published in the top journal in Political Science (*American Political Science Review*) and discovered that from 1970 to the 2000s, the percentage of studies that were based on causal forms of argumentation increased from only 10% in 1960 to over 80% in the 2000s (Gerring 2012, 730). A near-identical reverse trend can be observed in terms of the decline of descriptive forms of argumentation. These trends have only since been exacerbated. Yet with great promises of advancing social science have come great faults and many unmet expectations.

As a substitute or supplement to causal law-based explanations in social science, more than three decades ago, a call to turn to mechanistic explanations was made. Due to the acknowledging of prediction failure (also referred to as the simple failure to predict) in social science, Jon Elster put out numerous high-profile publications that called for turning to mechanisms as analytical tools in social science research. Elster argued that mechanisms can serve as a sort of intermediate function between laws and descriptions. He defined a mechanism as follows, "Roughly speaking, mechanisms are frequently occurring and easily recognizable causal patterns that are triggered under generally unknown conditions or with indeterminate consequences. They allow us to explain but not to predict" (Elster 1998). Elster put forward ideas about different types of mechanisms and situated his framework in the category of methodological individualism – a popular approach to social inquiry that assumes social phenomena can be explained in terms of individual units and their behavior (Elster 1998, 47). The ultimate aims of Elster and other proponents of mechanistic explanations (e.g., Mahoney and Rueschemeyer 2003) are twofold: first, proponents of these approaches seek to open up what many refer to as a "black box" of causality. If a causal inference is made after adequate statistical analyses are applied to a given set of data and a hypothesis testing inquiry, then mechanisms can help make sense of what connects a given independent variable to a dependent variable. Second, mechanisms enable researchers to reduce and disaggregate social phenomena–a task that some believe to be necessary in order to understand the complexity and lack of uniformity inherent to social reality.

As subsequent chapters of this book will demonstrate, the term mechanism is one that is heterogeneous across our lexicon. This discursive variance is not simply random, but symbolizes a deeply important tendency inherent to the history of science, causality, and human cognition. Mechanisms constitute a much grander and deeper set of ideas and processes pertaining social causality. Throughout human history, semantic alterations of causal terms have arisen according to changes in referents (physical changes in the world) and categorization (changes in world view). Prior to the onset of modernity, understandings of what constituted a causal force, for example, the mechanism, contrasted to understandings of these phenomena in both present day natural and social sciences. In the next chapter we will explore how the mechanism was once conceived as a theistic force that was God-like in character in Ancient Greece and an era where mysticism and metaphysics were at their historical peek.

References

Al-Daffa, Ali Abdullah. 2020. *The Muslim Contribution to Mathematics*. London: Taylor and Francis.

Anisin, Alexei. 2021. "Humanity and the Disruption of the Cosmos: How Berdyaev Foresaw our Reliance on Machines." *Cosmos and History: The Journal of Natural and Social Philosophy* 17 (3): 33–59.

Berdyaev, Nikolai. 1952. *The Realm of Spirit and the Realm of Caesar*. London: Donald A. Lowrie.

Bulmer, Martin, Bales, Kevin and Kathryn Kish Sklar. 1991. *The Social Survey in Historical Perspective. 1880-1940*. Cambridge: Cambridge University Press.

Cohen, Bernard. 1985. *Revolution in Science*. Cambridge: Harvard University Press.

Ebry, David. Ed. 2015. *Theory and Practice in Aristotle's Natural Science*. Cambridge: Cambridge University Press.

Elster, Jon. 1998. "A Plea for Mechanisms." In *Social Mechanisms: An Analytical Approach to Social Theory,* edited by Peter Hedstriim and Richard Swedberg, 45-73. Cambridge: Cambridge University Press.

Ewing, Alfred. 1969. *Kant's Treatment of Causality*. London: Routledge & Kegan Paul, Ltd.

Experts, Disha. 2019. *Epitome of Ethics, Integrity & Aptitude for UPSC Civil Services General Studies Mains Paper IV*. New Delhi: Disha Publication.

Gerring, John. 2012. "Mere Description." *British Journal of Political Science* 42 (4): 721-746.

Grant, Edward. 2007. *A History of Natural Philosophy: From the Ancient World to the 19th century.* Cambridge: Cambridge University Press.

Goudsblom, Johan. 1992. "The Civilizing Process and the Domestication of Fire." *Journal of World History* 3 (1): 1-12.

Gorton, William A. 2012. *Karl Popper and the Social Sciences*. Albany: State University of New York Press.

Guthrie, William Keith Chambers. 2012. *The Greek Philosophers: From Thales to Aristotle*. London: Taylor and Francis.

Henry, John. 2002. *The Scientific Revolution and the Origins of Modern Science*. New York: Bloomsbury Publishing.

Johnson, R, Monte. 2015. "Aristotle's architectonic sciences," In, *Theory and Practice in Aristotle's Natural Science*, edited by Ebry, David, 163-186. Cambridge: Cambridge University Press.

Kouremenos, Theokritos. 2018. *Plato's Forms, Mathematics and Astronomy*. Berlin: De Gruyter.

Kumar, Alok. 2014. *Sciences of the Ancient Hindus: Unlocking Nature in the Pursuit of Salvation*. Scotts Valley: CreateSpace Independent Publishing Platform.

Kristeller, Paul Oskar. 1964. *Eight Philosophers of the Italian Renaissance*. Stanford: Stanford University Press.

Ledoux, Stephen F. 2022. "Defining Natural Sciences." *Behaviorology Today* 5 (1)): 34-36.

G. E. R. Lloyd, 1970. *Early Greek Science: Thales to Aristotle*. New York: W. W. Norton.

Gordon, Scott. 1991. *The History and Philosophy of Social Science*. London: Routledge.

Lodge, Rupert. 2014. *Plato's Theory of Art*. London: Taylor & Francis.

Mahoney, James, and Dietrich Rueschemeyer. 2003. *Comparative Historical Analysis in the Social Sciences, Cambridge Studies in Comparative Politics*. New York: Cambridge University Press.

Marie Leroi, Antoine. 2015. *The Lagoon: How Aristotle Invented Science*. New York: Penguin Books.

Merriam-Webster. 2021b. "Natural science." In Merriam-Webster.com dictionary. Retrieved December 23, 2021, from https://www.merriam-webster.com/dictionary/natural%20science

Merriam-Webster. 2021b. "Social science." In Merriam-Webster.com dictionary. Retrieved December 23, 2021, from https://www.merriam-webster.com/dictionary/social%20science

Michels, Robert. 1911. *Zur Sociologie des Parteiwesen in der Moderned Demokrative*. Leipzig: Klinkhardt

Morton, Rebecca B, and Kenneth C. Williams. 2010. *Experimental Political Science and the Study of Causality: From Nature to the Lab*. Cambridge: Cambridge University Press.

Shapin, Stephen. 2018. *The Scientific Revolution*. Chicago: University of Chicago Press.

Sharpes, Donald. 2009. *The Evolution of the Social Sciences*. New York: Lexington Books.

Sutton, Daud. 2002. *Platonic & Archimedean Solids*. New York: Bloomsbury.
Van Melsen, Andrew. 2004. *From Atomos to Atom: The History of the Concept Atom*. Dover: Courier Corporation.
Watson, Peter. 2005. *Ideas: A history of Thought and Invention, from Fire to Freud*. London: Harper Perennial.

CHAPTER THREE

MECHANISMS IN ANCIENT GREECE

A calm sea touches the banks of sun scorched harbor during the autumn months in the Southern Mediterranean. Sailors are unloading their boats and lounging on the nearby deck while onlookers are taking their evening strolls. In the nearby distance, roars and gasps were heard from a crowd. A light almost as strong as the setting sun emits an evening gloom from an Athenian theater–a place perched on a hillside filled with an audience surrounded by an altar, orchestra, and skene. A Greek tragedy is being enacted and like all Greek tragedies, it features a serious and substantial display of artwork in which the relationship between life and death, between cause and effect, and the nature of suffering is explored. This is how evenings in Athens transpired at the height of Ancient Greek civilization, an era that was dominated by the advent of philosophy, metaphysics, and tragedy. The latter was, in its essence, a work of great aesthetic pleasure (Hall 2010, 11). Greek theater experienced its heyday during the 4th and 5th centuries BC–a time marked by the emergence of the polity and first democratic social order in the history of humanity. Wright (2008) describes the special nature of Greek theater as follows, "It is amazing that, despite the huge cultural differences that exist between our world and that of the characters, we can still respond with such intensity to the tragic myth" (Wright 2008, 51).

It is within this prominent historical era that we begin our investigation of the origin of the mechanism and its emergent role in symbolizing how human populations conceived of and interpreted social causality. The context of Ancient Greece is not only known for its forging of the first democratic polity, the invention of philosophy, nor of the development of scientific thoughts and experiments. Ancient Greece, above all, was a civilization that efficaciously merged polytheistic beliefs into metaphysical narratives about causality. In the previous chapter, I identified the monumentally significant contributions to knowledge, philosophy, and culture that the Ancient Greek civilization put forward. This chapter will delve into the specific social and anthropological settings that underpinned Ancient Greek society. It was in these particular settings that the first conceptions of the mechanism arose. This chapter will then overview and identify the usage of mechanisms in some of Ancient Greece's most noteworthy tragedies. The analysis will enable readers to situate the linguistic, symbolic, and metaphysical parameters of the topic under attention.

All usages of the term mechanism in the English language (as well as in Russian, German, along with Latin-based languages) originate to the Greek term μηχανισμός (mechanism) or μηχανή (michani). In the Greek language, the discursive function of the term mechanism is widespread (as it also is in other languages, including English), and is synonymous with objects such as machinery, tools, a piece of equipment, movement, machine, and device. A fascinating dynamic here can be observed in the fact that the etymology of the Greek term for mechanism, μηχανισμός, is in theater and the usage of a machine (mēchanē) that was implemented during actual theater shows to alter the plot of a story or radically transform the entire story itself. Because of its importance to theater, later on the Roman and Latin-speaking world adopted the semantic interpretation of the μηχανισμός as the deus ex machina (God from the machine). According to the Merriam-Webster dictionary, the deus ex machina is a noun that has a twofold definition: 1) a god introduced by means of a crane in Ancient Greek and Roman drama to decide the final outcome, or 2) a person or thing (as in fiction or drama) that appears or is introduced suddenly and unexpectedly and provides a contrived solution to an apparently insoluble difficulty (Merriam-Webster 2022). The deus ex machina functioned as a plot shifting story and arguably the first usage of this concept was in Homer's *Illiad* and the ending part of the epic of sailing home (Johnston 2019, 125). Johnston (2019) argues that there is little evidence that the mechane and deus ex machina were used before the tragedy of Hippolytus (428 BC) (Johnston 2019, 128).

An important fact to take into consideration while reading this chapter (and the subsequent two chapters), is that not every single tragedy featured the implementation of a deus ex machina. Some tragedies had plots that were shifted or resolved in quite normal and non-terrific ways. But there also were many tragedies that did indeed rely on the deus ex machina, and some of these tragedies are considered by audiences to be the most significant of all Ancient Greek works. Most significantly, the usage of the deus ex machina in Greek theater served as an exemplar of understandings of social causation of this era of history. During theater shows the deus ex machina was materialized in the creation of a lifting machine that would enable objects (which often were sculptures of gods, god-like characters or heroes of a story) to be raised and to hover over a stage from a range of four to eight meters (Dimarogonas 1992; Papadogiannis et al. 2010). The mechane comprised wooden cranes that were used throughout a given spectacle. Often, a lever that was high-above stage would swing down to bring characters directly onstage or through a hidden trapped door (Ashby 1999). There was even a scientific field where the construction of such mechanisms in Ancient Greece took place and was referred to as μηχανική. The machine would function through a pulley that was on both ends of its wooden beam. On the bottom of the wooden beam were metal weights that would serve to counter the load that would be risen. This theatrical tool was used alongside other feats of mechanical engineering including the periatkos–a revolving block that would enable quick changes of scenery to occur in a play, along with the ekkyklema–a wheeled vehicle that would carry heavy objects to and from the stage (Chondros 2004, 89).

Interestingly enough, prior to Greek theater, the mechane had no forerunners in other engineering contexts. As described by, Chondros (2004), "the mechane does not appear to have any close predecessor and its development is of great engineering significance since it appears that it was designed to meet very close requirements of the play and it was not arrived at by long evolution (Chondros 2004, 89). Hence, the creation of the mechane and its usage in theater was not a random development nor was it an intuitive linguistic categorization of an important cultural phenomenon, but rather, the deus ex machina signifies a number of different elements that were prominent to human cognition at an early point in our historical development as a species. Before unpacking these complex dynamics, in-depth inquiry must be given to the context of Ancient Greek theater. When we hear the term "theater" today, we often identify it with a niche of society–many know that there is a sub-set of society that enjoys going to theater shows, and theater has a solidified place in most of Western culture (as well as many Eastern cultures). Yet in the Ancient Greek context, theater was all encompassing of culture–it signified the height of civilization and was among the only forms of public entertainment. This era of history was also marked by a larger process that humanity was entrenched in. Berdyaev referred to this age in his periodic classification of humanity's relationship to nature as the first period, a period that was based on our submersion to cosmic life, "in which human life depended on the natural world–a time when personality was not fully developed and humans did not fully conquer nature" (Anisin 2021, 35). While some of the most celebrated endeavors of philosophizing can be traced to great Ancient Greek figures such as Aristotle and countless others, this era of history was dominated by polytheistic conceptions of life, and the average civilian (not to mention slave), was far removed from philosophy. Yet the average civilian was also present social life wherein theater and storytelling practices were prominent. Here, great tragedies would be told verbally in communal settings across generations.

The conception of a mechanistic force was contingent on the polytheistic world views of this era. During this time period, humans were still submerged into cosmic life and their relation towards nature was based upon magic and myth (Berdyaev 1952, 47). In the Ancient Greek era as well as preceding eras of human history, natural elements and phenomena were conceptualized by populations as being theistic in character. For instance, Watson (2005) identifies the concept of a "sky god" to have arisen in early Neolithic times. In Mesopotamia there around 3,300 different names for gods and the Sumerian word for divinity was analogous to "bright" and "shining" (Watson 2005, 101). What's more, In Aryan tribes, the God Diues, was the God of the light sky which brings us to a similar deity, the Greek God, Zeus, who was also believed to have evolved from a primitive sky divinity. What's more, these early time periods of human history featured articulations of different conceptions of the soul and how it relates to a possible afterlife. The importance of these cognitive developments lays in ideas about what constitutes heaven, hell, and good and evil. According to some, the Greeks were the first to articulate the concept of an immortal soul due to ideas put forward by noteworthy figures such as Pythagoras who thought that souls were in themselves gods and were imprisoned in human beings' body (as if in a tomb) and were heading towards rebirth. Other contributions can be observed in the works of Parmenides, Herodotus, and Empedocles (Endsjø 2009, 106).

The Ancient Greek conception of an immortal soul came at a time (750-350 BC) in which most of the world's greatest faiths formed (Watson 2005, 107). For our purposes, the salience of this historical period can be observed in the fact that populations who articulated and held on to these faiths did so via the attribution of theistic powers to nature and the phenomenal world. Indeed, these were ritualistic attributions because human faces and names were associated with natural forces in ways that revolved around natural events (hurricanes, floods, storms, earthquakes, etc.,) and such events were widely revered by populations. Furthermore, the emergence of theater came about at a time of prosperity in Ancient Greece and an age where Dionysus (the God of the vine) was widely worshiped. In rituals, citizens would gather together to worship Dionysus and a goat would be sacrificed. As Watson (2005) points out, there was a direct link between sacrifice and tragedy because this "primitive ritual lives in our most powerful form of theater" (Watson 2005, 139). Interestingly enough, the word tragedy has etymological roots in the Greek "trag-odia" which meant religious celebration that was narrated by a single person called the Responder. The Responder was supported by a chorus who sang and danced. As time went on, narratives expanded and gods other than Dionysus were represented in such rituals alongside robust and attractive dialogues. This led to a common understanding that a voice that would represent different characters and a new complexity in narratives. At annual festivals, the tragic drama was established and prizes were given for technical innovations that were used to support these shows (Watson 2005, 139). The festival of Dionysus was a major public event that would last over a week and feature a combination of religious rituals, ceremonies that were political in character, along with a variety of musical and dramatic performances (Wright 2008, 8).

In contrast to other traditions and faiths of that era, the Ancient Greeks were undoubtedly the most entrenched in rationalist ideas as they combined rationalism with metaphysics to form arguably the first coherent framework of natural philosophy (as described in the previous chapter). This is likely why the actual physical creation of a mechane was made through implementation of mechanical engineering and emerged within Greek theater. The mechane was a natural offshoot of rationalizing concepts from the noumenal realm into the phenomenal. As I have argued thus far, a fruitful way to simplify this overly complex set of dynamics is to consider the Ancient Greek advent of a mechanism or the God from the Machine through a Kantian lens. The notion of a mechanism represents the synthetic a-priori abilities of the human mind in being able to stipulate a cause. At this early period of history however, the character of the mechanism was polytheistic and attributed to predetermined conceptions of the gods. What arose from this was the physical construction of a mechane–a tool that was made out of wood and hovered on top of or behind the stage of a given theater. This element of rationalism in early Greek thought is of great importance to both the physical creation of a mechane as well as its cognitive articulation of a force that has causal propensity. The God from the machine (deus ex machina) exemplified divine will and was monumentally important to how populations in this context conceived of historical change as well as changes in their own personal lives and societies.

A contradiction may appear to readers in that this polytheistic dominated era of culture and religion was also one that bred the first advents of comprehensive forms of scientific inquiry. How can this major discrepancy be accounted for? Why do general societal tendencies and beliefs tend to lag behind the innovative discoveries that get made by particular individuals or groups? As subsequent chapters will reveal, this tendency is common to all periods of human history under attention in this book including–the Middle Ages, the industrial era, and the contemporary era of digital technology and artificial intelligence. In each of these eras, philosophical ideas and frameworks as well as rates of scientific progression always surpassed the common plane of cognition shared by the general population. This is not meant to be a derogatory statement or assumption about the intellectual status of human beings, but is one that seems to be endemic to the human cognition–in that ideas and inventions always arise, get articulated, and get created, but very few people are aware of the implications or nature of these inventions until decades and sometimes even centuries later. This is why even in light of essentially discovering philosophy, mathematics, physics, and chemistry (among other topics), Ancient Greek societies still attributed mechanistic activity to theistic forces–forces that were representative of natural phenomena with a human face. No better example of this can be found than the great Greek tragedies written by figures such as Sophocles, Eurpides, Aeschylus–who in total, created around two to three hundred tragic dramas (Hall 2010).

Greek Tragedy and Athenian Theater

Athenian theater is a term that is interchangeably used to describe Greek theater as a whole because this is where nearly all Greek tragedies were enacted. Estimates of the total population of Athens at height of Greek civilization vary and the average appears to be around 30,000 citizens and hundreds of thousands of slaves, many of which were agricultural laborers (Roselli 2011, 14). Within Athenian theater, audience interactions constituted political actions, argues Roselli (2011), and the space as a whole functioned as a zone where ideas were produced about social relations and the community. Hall (2010) notes that Athenian theater fostered the production of identities for the population. Geospatially, the theater at Dionysus (in Athens) had an arrangement that featured more seating capacity than other theaters of its time. In open air, gatherings were essential to shaping group opinions (Hall 2010, 60). Crowds at the theater were in themselves an exemplar of public opinion and democracy. People would shout, boo, whistle, hiss, bang their feet, and clap. This would all lead to different output of a mixture of emotions and rationale, of median positions and radical views–all of which would fuse into a socio-cultural cocktail that had positive effects on political order.

These tragedies presented audiences with inquiries into suffering through as aesthetic questions were posed on topics of pain and suffrage (Hall 2010, 6). As eloquently described by Hall (2010), Greek theater and its tragedies were based on the relationship between the actual theater and the audience, "Tragic audiences shared their humanity—their vulnerability to suffering and their inevitable mortality—with the heroic characters the wrecking of whose lives they were witnessing. But as elevated spectators they had more in common with the 'internal audience' of gods who caused that suffering" (Hall 2010, 8). As such, the inclusion of theistic forces into tragedies and the immortality of the gods was done for the purpose of getting spectators to experience and comprehend ideas on immortality. Regardless of how powerful a given character in a given tragedy may have appeared, his/her ultimate power was limited in light of the immortal God. Likewise, as argued by Hall (2010), even though a given spectator may have known much more about the characters than themselves, the spectator still had absolutely no power to intervene because of the luminescent supremacy of the gods. This made spectators moral witnesses but not moral agents (Hall 2010, 8). Along these lines, scholars of Greek theater acknowledge the salience that theistic forces played in shaping story lines and cultural scripts. For instance, Wright (2008) notes that, "In particular, the fifth-century Greeks would have found it hard to conceive of such matters as character, motivation, responsibility, or normal or abnormal states of mind on a purely human level, without reference to the gods" (Wright 2008, 52).

Some of the earliest of great dramatists enacted their own tragedies verbally in small theatrical settings. For example, Thepsis (6th century BC) is considered by some as the Father of Greek tragedy and was a poet. He was a part of a group of some of the earliest playwrights which would go onto stages and play roles from their own plays. A similar example can be observed in Aeschylos, a war hero that later won literary contests at drama festivals in Athens (in the year 484 BC) (Chondros 2004, 87).

This brings us to our first great tragedy–Euripides' Medea. Euripides (480–406 BC), a Sophist, is historically renown for famous tragedies including Medea, Hyppolytus, Electra, the Trojan Woman, among dozens of others. His work is widely regarded as being among the most characteristic of Greek tragedy and his versatility can be observed in the fact that a number of his stories were also political in nature (e.g. the

Suppliants) as well as romantic (e.g. Ion or Helen). Euripides has a particular ability to capture how human personality could deteriorate as a response to vengeance (Watson 2005, 140). Along with Medea and Orestes, Euripides utilized the deus ex machina in the following plays: Andromache, Bacchae, Electra, Hippolytus, Suppliant Women, Iphigenia among the Taurians, Helen, and Ion and Rhesus (Wright 2008, 49).

Euripides' Medea

Medea is particularly important because its conclusion is resolved by usage of a deus ex machina and as a whole, Medea is one of history's most noteworthy stories and plays. Some scholars argue that Medea is the first play to have incorporated the usage of a deus ex machina. The play does not represent a conscious echo of preexisting scene-types, but rather is an "exploration of the dramaturgical potential that the mechane offered before its conventional use to introduce gods at the ends of plays had become established" (Johnston 2019, 126). First performed in 431 BC in front of 15,000-20,000 spectators in Athens (Allan 2002, 12), Medea emerged just several months prior to the outbreak of war between Athens and Sparta. Allan (2002) contextualizes the play in the following way, "The play's contrast between a Corinthian king who expels an innocent woman and an Athenian king who agrees to protect her is no doubt significant" (Allan 2002, 17). Entrenched in a conflict-ridden set of circumstances, the story takes place in the city of Corinth and revolves around the province's King, Creon, his daughter (Glauce), and Jason (the same Jason from the story of the Golden Fleece and Jason and the Argonauts). Along these lines, the main character of the story, Jason, leaves his wife Medea and their children in attempt to coalesce with a richer and more socially prestigious woman–Glauce. Medea plots revenge against Jason for his betrayal and knavery. Not only does Medea curse Jason's life, but she curses her own life as well as her children's. Sensing her instability, King Creon bans Medea from Corinth. He refers to Medea as follows, "You there, with your scowling face" (Allan 2002, 28).

Meanwhile, Medea is plotting to achieve justice on her own through killing all parties involved–Creon, Glauce, and Jason. While on tour in Corinth, a King from Athens, Aegeus, observes the turbulent situation and offers Medea exile in Athens in exchange for her knowledge on sorcery and critical medicinal properties that the king believes can help his personal ailing condition. Medea interprets this as a chance carry out her plan of revenge and then escape. She then coerces Glauce to accept

Jason's children into her inner-circle and to have them stay in Corinth. By doing this, Medea offers Glauce a golden coronet that she secretively poisoned. This leads to Glauce's death–a death that was portrayed as being exceptionally gruesome and even heinous. A messenger in the tragedy accounted for the death to audiences by stating it devoured the woman's flesh and "set her head on fire" (Allan 2002, 40). Further, Creon arrived shortly after and witnessed this set of events and then took his own life. In a shocking turn of events, Medea also murders her own children through the same poisoned garments.

After Medea announced she would kill her kids, audiences were reportedly shocked and wondered whether anyone in the story could actually stop Medea. This built suspense towards the horrendous climax of the play (Alan 2002, 35). The importance here is in the particular climax that followed after Medea's announcement. A climatic set of events enabled this tragedy (like many others after it) to flow from its second to third act with great suspense. Allan (2002) describes Euripides' effectiveness in being able to achieve powerful effects through the usage of emotional suspense, disturbing violence, and dramatic irony. When Jason arrived to the palace to see what had transpired, he stated the following about Medea, "She must either hide herself beneath the earth or soar on wings into the heights of the sky" (Allan 2002, 42). The latter is what ended up happening as Medea fled from Corinth in a chariot (made possible via the deus ex machina) hoisted by a mechane (Worthington 1990). She escaped via a chariot that was sent by her grandfather and sun-God Helios. The usage of the deus ex machina here exemplifies the supreme force of the gods, and at the time, it offered audiences a metaphysical solution to an engrossing story. The deus ex machina dominates the final scene of this tragedy and signifies a distance between both human and divine approaches to justice, argues Allan (2002, 94).

The end of this tragedy is marked by Jason grieving over the entire conflict, the loss of his children, and the death of Creon. All of Jason's personal relationships and life goals are shattered and his code of honor is broken. Johnston (2019) notably argues that mechanistic usage here is not exemplary of other comparative implementations of the deus ex machina because Medea does not enter divinity herself and is only hoisted by a god-sent chariot. In contrast, Allan (2002) interprets the final scene as Medea possessing God-like authority because "no ordinary Athenian woman could so punish her husband and escape" (Allan 2002, 98). Allan also describes Medea's killing of Glauce and her own children as actions that "combines Medea's god-like impunity with her human pain, and her ruthless punishment of her enemy with her self-destructive, and deeply tragic, revenge" (Allan 2002, 99). The latter is, in my view, the overarching point of the deus ex machina in that it manifests itself as a heterogeneous plot shifting force, and one that possesses power from theistic forces that function outside the realm of earthly materials. Its origin is outside of our sensory experiences–an origin that is best summarized as a derivative of the noumenal world. The usage of the deus ex machina enabled populations to make sense of causes and effects.

Euripides' Orestes

The next tragedy under attention is also written by Euripides–Orestes (first performed in 408 BC). This is his last written tragedy and in contrast to the previous tragedy, Orestes also is much more embedded in institutions. A trial along with legal constraints dictate large portions of the story. This play also features the presence of theistic forces to a much greater extent when compared to Medea, especially as observed in the actions of the God Apollo. Apollo was considered by the Greeks to be a multifaceted God and undoubtedly one of the most important–he was God of music, dance, truth, prophecy, healing, disease, and archery. The main character of the story is Orestes (son of Agamemnon and Clytemnestra) and his sister Electra. With aid of Electra, Orestes kills his own mother out of revenge because she had killed his father after following Apollo's advice. Before delving into the deeply rooted metaphysical forces present in this story, we must consider that the play was written and took place after other plays written by Euripides including Eltra and Helen which makes it apart of a trilogy. The play was also written and enacted during the Peloponnesian war which was one of the longest enduring conflicts in human history.

The opening segments of the story are marked by Orestes facing the death penalty for murdering his mother. The only scenario presented to audiences of his potential to survive is presented as help from Menelaus who arrives back to Athens with Helen. During the early stage of this tragedy, the prologue is dominated by Orestes' anguish which is instilled on him by a goddess, the goddess of revenge, Erinyes–whose forces make the main character hallucinate and lose his self-awareness. During these hallucinations, the first entrance of Apollo is witnessed when Erinyes is attacking Orestes. Orestes asks Apollo for his bow to protect him from the goddess of revenge. Later on, when Menelaus arrives, Tyndareus (the father of Clytemnestra) makes a riveting speech that calls for Orestes' execution for his crime–referring to Orestes as follows, "Ah! What an evil it is to be ignorant of the future! There is that matricide before the house, a viper darting venomous flashes from his eyes, whom I loathe. Menelaus, are you speaking to that godless wretch?"

In speech, Tyndareus threatens Menelaus with exclusion from Athens, yet Menelaus still sees merit in Orestes' defense because it was Apollo who convinced him to kill his mother. This aspect of the tragedy exemplifies the role of theistic powers in determining character behaviors and decision making. After the onslaught brought forward by Tyndareus, Orestes is left with little hope and thus ponders over his impending death. Sometime later, one of his close friends named Pylades returns and convinces Orestes to defend himself. In an uninspiring defense, Orestes does not succeed and is convicted of death by way of stoning (along with Electra). Hearing this verdict, Orestes speaks to the assembly in attempt to convince them of letting him and his sister have the chance to take their own lives rather than face a harsh execution by getting stoned to death. After leaving the assembly (the court house of the time), Pylades conspires a new plan that he believes will give Orestes a better legacy after his death–to kill Helen, the wife of Menelaus and a figure who was bereaved in their societal context. Both Orestes and Electra agree to this plan and Electra also proposes they take one of Menelaus' children (Hermoine) hostage so they can coerce him to rescue them. In the fifth act of the tragedy, this plan moves forward with Electra vehemently leading the way. Helen is apparently killed and then Hermione is captured by Orestes and Pylades. While this scene was enacted, a scream was depicted as being Helen's, yet her actual physical status was never revealed to audiences. After the scene ends, a slave emerges from the roof of the building that these interactions were transpiring in and reveals to the audience that Helen vanished in thin air and managed to escape through divine intervention. This scene depicts what Wright (2008) describes as "one of the most astonishing entrances in Greek tragedy" (Wright 2008, 45), because audiences expected that Helen's corpse would emerge instead of the slave.

Meanwhile, citizens of the city of Athens are in an uproar after hearing what is transpiring in the palace which results in Menelaus rushing in to save his daughter along with soldiers and members of the public. Once on the scene, Menelaus orders the soldiers to break through the palace doors but is immediately met by Orestes who emerges onto the rooftop holding a sword to Hermoine's throat with Pylades next to him with lit torches. Just as Orestes is about to slice Hermione to death, both Helen and Apollo come forth on top of the roof (via the mechane) in an exemplary manifestation of the deus ex machina. Apollo speaks and describes how

Helen was brought to heaven because of Zeus' will and simultaneously, Apollo calls for this vengeful conflict to end. He then states that Orestes will be exiled for some time, but then will come back to Athens and be acquitted for his crimes and will subsequently be appointed a King of Argos then will marry Hermione. The Chorus of the theater sings in celebration and the tragedy comes to an end. This final scene is representative of a configuration that is spectacular and exemplary of Greek tragedy and theater because "the orchestra, the front of the skene, the roof, and the mechane on high are all full of activity (Wright 2008, 49).

It is clear why historical and political importance have long been attributed to this play. Apollo's call for peace came in a time of Greek history that was marked by the Peloponnesian conflict (Zuntz 1955). For our purposes, we must consider that the usage of the deus ex machina is also of outstanding significance because like many other great tragedies, it determined the outcome of this tragedy.

Aristophanes' Peace

Alongside Medea and Orestes, another example can be observed in a what many consider to be an optimistic and pleasant comedy called "Peace" (*Eirene* in Greek). Peace was written by Aristophanes and was enacted in Dionysia just before the Peace of Nicias (421 BC) which was an attempt to end the Peloponnesian war 10 years into conflict by bringing about a truce. Here a chief protagonist, Trygaeus, searches for answers that would potentially aid him in being able to put together a plan to end the war. Trygaeus is portrayed as a regular Athenian citizen who sets out to resolve the longest conflict of his time through searching for a lost goddess called Peace. The deus ex machina gets utilized by Aristophanes very early on in this story compared to the previous tragedies discussed in this chapter (where it was used in epilogues). The first scenes in Peace feature slaves working outside in the country side of Athens, digging up large pieces of dirt that they would then feed to a monstrously large beetle. While digging, Trygaeus flies by on the back of the beetle. In theater representations of this scene, the main character got hauled up by a mechane above stage while riding a large make-shift beetle. Below, characters playing his children and neighbors were pleading for him to return back down, claiming he did not know what great dangers may lay ahead as he ascended up to the sky, to the realm of the gods. Yet Trygaeus defended his actions by stating he was riding the beetle to go speak to the gods in attempt to end the war. Similar to Medea, the usage of the deus ex machine is to propel a human being up into the stratosphere of the gods and attempt to engage with their domain.

While flying by the slaves, Trygaeus made a compelling speech to those below him in which he argued that he sought to reason and bargain with the gods to end the war. This premise to the overall story of Peace once again exemplifies the societal views and beliefs of the Ancient Greeks to forefront–causation could only be determined and made possible by the will of gods. Moreover, when Trygaeus arrives into the heavens, he meets the God Hermes who reveals that all other gods have left and that the God called War imprisoned the goddess Peace in a pit. War then emerges with intent to kill all of the Greeks. Trygaeus, using his last possible hope, goes to summon his fellow countrymen to attempt to free Peace from the pit which is situated in a large cave. As the Greeks pull large rocks from the pit, it is clear that those whose occupation is farming are the most successful in the endeavor. They finally reach Peace and set her free along with two others whose names are Harvest and Festival. The goddess then explains her rationale for leaving years earlier–she was credibly fed up with the political actors that were receiving material profits from waging war. What's more, Hermes states that it was the people of the Athens Assembly that voted against her release on multiple occasions.

Shocked by these revelations, Trygaeus offers a heartfelt apology to Peace and then leaves to go back to Athens alongside both Harvest (whom he later marries) and Festival. Once back, a pleasant Chorus offers praise for Trygaeus' actions and roasts a lamb as a sacrificial ritual prior to his wedding. Although Trygaeus did end up successfully ending the Peloponnesian conflict, he importantly sided with regular common people and went against a class of traders and political actors that were depicted as warmongers because of their repeated profiteering from the war. This contrasts with another main character in the story in Athens' leader, Cleon, who is depicted by Aristophanes as being an authoritative war mongerer. The epilogue of Peace emphasizes a degree of collective solidarity that is shared by farmers and those living in relatively simple forms of socio-economic organization in the countryside. For our purposes, Peace offers a glimpse ahead into how other playwrights' centuries down the line also utilized the deus ex machina in periods of their stories that fall outside of the epilogue. The notion of flying away on a large insect, animal, or mythical creature was later utilized by authors across some of the most prized stories and novels that date to the Middle Ages.

What the Deus ex Machina tells us about Social Causality

The mechane helped to incorporate divine forces into a given literary or theatrical composition. It was utilized to integrate the act, intention, and supremacy of various gods into stories. It played a crucial function to solve a problem in a story or to resolve a plot in its entirety. As this chapter has demonstrated, prominent examples of mechanistic plot shifts can be found throughout Ancient Greece's most significant theatrical tragedies. This ancient context was not randomly chosen to assess how populations conceived of social causation, but was selected because due to it being the origin of the mechane and the deus ex machina. This context also exemplifies an era of human history in which human beings' a-priori synthetic knowledge was just starting to develop. As noted in the previous chapter, the Greeks put forward a number of monumentally transformative advancements to knowledge through the creation of philosophy (which would manifest as natural philosophy for the next near-two thousand years), mathematics, and many other technical and cultural innovations discussed thus far. The combination of the deus ex machina with the physical creation of a mechane signifies how conceptions of causality stemming to the noumenal realm were complimented by the usage of a-priori knowledge and then manifested into a technical tool in the phenomenal realm.

In Kantian terms, the historical creation of a mechane was an attempt to represent the principal source of understanding of power and causality that people held of the noumenal world. Since the noumenal world is comprised of God, essences (metaphysical realities and substance), and the self, we cannot fully grasp the nature of forces that we conceive to be God-like or supernatural. As such, in order to mentally forge the mechane (i.e., to conceive of its design before its actual physical creation), a-priori knowledge was drawn on and resulted in an attempt of displaying metaphysical essences into the real world. This included understandings of space and time–the mechane needed to be placed in the right area of a theatrical composition (hovering on the top of a stage) and it needed to function in a way that would enable it to be deployed and holster characters are precise segments of a story. The creation of the mechane was a partial attempt by the Ancient Greeks to recognize what Kant believed to be unknowable–gods and supernatural forces, both of which Kant assigned to the noumenal world. This exemplified societal views on causation that were not only limited to the Peloponnesian part of the world, but rather, were representative of humanity as a whole.

With that being said, even though the Ancient Greeks fostered a monumental alteration in the development of human beings' ability to engage in rationalization of the natural world, apart from scholars and members of society that were "learned," it would take almost two thousand years for another revolution of this sort to arise with the emergence of new ideas put forward by figures such as Copernicus and Newton. The Greek noun *aitia* and its adjective *aitios* and *aition* is translated as "cause" and some refer to it as "reason" or "explanation" (Broadie 2012, 21). Broadie (2012) points out that the early Greek usage of aitia was used to refer to a certain property or status (X) with relation to Y, and that, "it is not simply to claim that X stands to Y in the relation of having done or brought it about, although this is implied" (Broadie 2012, 21). Eventually, these terms were applied in theoretical inquiries in wider systematic explanations in which assigning or ascribing X to the status of *aitia* of Y was now done to understand Y in a scientific way (Broadie 2012, 22). Broadie argues that it was not accidental that the meaning of the word *aitia* brought about debates in Ancient Greece based around the causes of things and about what counts as explanation. Through such debates, Aristotle argued against Plato's forms and shifted attention into an ontology of objects.

Along these lines, as I have tried to stress throughout this chapter, even with very innovative developments to natural philosophy and early scientific inquiry being made at the height of Ancient Greek civilization, societal tendencies took place in a context that was marked by polytheistic understandings of the world. It may well be the case that our cognitive capacities are not static and are historically evolutionary. The a-priori knowledge utilized to create the mechane was impressive by any standard, yet it is technically infantile when compared to the inventions brought about by the scientific revolution. Since our a-priori abilities appear to be historically evolutionary in their character, this lends support to the general claim that what we conceive to be causation, namely social causation, is also evolutionary and not static.

Furthermore, while scholars have identified the disconnectedness that characters in Greek tragedies shared or experienced with relation to the overall outcomes they endured, these particular dynamics have not, to the best of my knowledge, been related to the ontological status of the Greeks' a-priori knowledge. Wright (2008) argues that in most Greek tragedies human characters do not actually understand what is going on because of deities and supernatural forces that are operating behind the scenes (Wright 2008, 64). In turn, this results in characters "trying, but failing, to interpret their lives in a broader context" (Wright 2008, 65). This claim is interesting to ponder over because it implies that the larger societal context in which these tragedies were enacted in was similarly unaware of material forces and their impact on physical and ideational outcomes. In Orestes, the main character functions as an agent of a God (Apollo) who encourages him to murder his mother. The fascinating dynamic here is, as Wright (2008) points out, that even with this encouragement, nowhere in the play does Euripides reveal why Apollo forced Orestes into doing this. The relationship of Orestes with a God is not close nor privileged (Wright 2008, 67). Similar issues can be observed in the role that Zeus has in Orestes and the God's decision to make sure Helen escapes and lives.

Another fascinating dynamic here can be observed in Apollo's final appearance in Orestes via the deus ex machina. His appearance is complimented by the presence of Helen (who was there because of Zeus' agency). One can pinpoint rationalist-based critiques of the deus ex machina that gained a degree of prominence at the time period of its inception. Aristotle directly criticized playwrights' usage of the deus ex machina throughout different sections of his *Poetics* (Baxter and Atherton 1997), by arguing that the mechane represented irrationality in tragedy. Since the mechane was not intelligibly used or mentioned as a component of character's knowledge of their world before the play, then its usage during a story is simply irrational. Shipley (1953) describes Aristotle's criticism of the deus ex machina as being marked by a sense of being unnatural because "unravelling" a given plot must arise from the actual plot itself and not some irrational external scope that is outside of the tragedy. In his criticisms, Aristotle was directly addressing theologians and the dominant viewpoints of his societal order that believed philosophical problems could be solved by means of the gods through constructing outcomes based on mysteries that could not be penetrated (Bodeus 2020, 217).

Whether intentionally or unintentionally, through his engagement with the deus ex machina and the semantic meaning(s) of the mechane, Aristotle sought for general knowledge of change and phenomena in the world to be conceived of through more specific and rational parameters. His criticism is important because it demonstrates a very early attempt to re-imagine the meaning of the deus ex machina.

Semantic Change

The mechane was conceived of, constructed, and utilized by the Greeks to enact the deus ex machina and from this period onward, these characteristics remained constant. It took many centuries for semantic changes to arise. By the time that natural science was starting to become solidified in a newly industrialized Europe, renown philosophers such as Schlegel made similar comments about Greek tragedy (and Euripides in particular) as Aristotle's. Schlegel argued that Eurpides' work and its dependence on the deus ex machina was of a lower intellectual level and less aesthetic than other tragedies (Micheleni 1987). Yet a lot happened from the point of the mechanism's inception until Schlegel's era and criticism of the deus ex machina. To date, no study or academic work has identified the trajectory of semantic changes associated with the mechanism. In the broad field of linguistics, semantic change of linguistic terms has been observed to be brought about by a heterogeneous collection of factors including those that are linguistic, psychological, socio-cultural, or encyclopedic in origin (Blank 1999). By semantic, we refer to an assumption that phenomena associate with meaning in a language or topic– "of or relating to the meanings of words and phrases" or "of or relating to semantics" (Merriam-Webster 2022b). Historically, the groundwork for theory on semantic change was established by Greek and Roman grammarians (Traugott and Dasher 2004, 51). By the 18th century dictionary making and comparative linguistics gained greater prominence and more attention and sophistication went into understanding semantic change (Traugott and Dasher 2004, 51).

This chapter has identified the semantic origin of the term mechanism and has analyzed its accompanying socio-cultural context of Ancient Greek theater. In an account of the impact that Greek tragedies had on their society, Sourvinou-Inwood (2003) argues that tragedies symbolized and reinforced their audiences' religious existence and realities. This reinforcive dynamic is what enables the semantic characteristics of a given term to hold their meaning across a given duration of time. When semantic change does arise, it has to involve observable evidence "of transmission of innovations to others" or what scholars refer to as conventionalization (Milroy 1992; Traugott 2014). Readers must understand that the study of semantic change emerged in the 20th century and has since seen great development throughout scholarship. In this academic field, scholars have identified different types of sub-categories of semantic changes including metaphorization (the conceptualization of one thing in terms of another); metonymization (how terms represent association, e.g. board table or keel for ship); pejoration (how terms associate with negative meanings); amelioration (how terms associate with positive meanings); narrowing (how words get restricted of meaning); generalization (how terms extend their meaning) (Traugott 2014).

This book does not place focus on a specific type of sub-category of semantic change because its exploration is not only in the linguistic term–mechanism, but is also embedded in the conditions that surround what the mechanism represents. This makes the analyses presented in this book semasiological. Semasiology is a linguistic approach used to investigate the meaning of a word independent of its phonetic expressions. This term has origin in the Greek term semasia (*meaning*). In contrast, the onomasoiological, Greek origin–*onoma* (name), has to do with studying how designations get made to a given object or concept (Traugott and Dasher 2004, 25). As the subsequent chapters will reveal, the designations made to the mechanism have all surrounded and revolved around causal forces, but the meanings that were attributed to the term throughout different historical eras changed according to whether the term was influenced by ideas about the noumenal world or the phenomenal world or a mixture of both.

If one were to open up a modern textbook on semantic change (or any other social scientific disciplinary topic) such as Traugott and Dasher's *Regularity in Semantic Change* (2004), they would quickly observe that discursive usage of the term mechanism is much different than the Ancient Greek interpretation of what constitutes the mechane. In books and studies on semantic change entire sections can be found that explore the "mechanisms of semantic change," just as in other disciplinary topics that explore a vast range of causal relationships and outcomes. The modern understanding of what constitutes the meaning of a mechanism has evolved into a highly empiricist and rationalist conception and one that Aristotle likely would have been very proud of. Before this stage in semantic change was reached however, understandings of mechanisms in the Middle Ages still were rooted in metaphysical underpinnings and it is this context that we will turn to next.

References

Allan, William. 2002. *Euripides: "Medea."* London: Brill Archive.

Anisin, Alexei. 2021. "Humanity and the Disruption of the Cosmos: How Berdyaev Foresaw our Reliance on Machines." *Cosmos and History: The Journal of Natural and Social Philosophy* 17 (3): 33–59.

Ashby, Clifford. 1998. *Classical Greek theatre: New Views of an Old Subject*. Iowa City: University of Iowa Press.

Baxter, John, and Patrick Atherton. 1997. *Aristotle's Poetics*. Montreal: McGill-Queen's University Press.

Berdyaev, Nikolai. 1952. *The Realm of Spirit and the Realm of Caesar*. London: Donald A. Lowrie.

Blank, Andreas. 1999. Why Do New Meanings Occur? A Cognitive Typology of the Motivations for Lexical Semantic Change, in *Historical Semantics and Cognition*, edited by Blank, Andreas, Koch, Peter, 61-90. New York: Mouton de Gruyter.

Bodeus, Richard. 2020. *Aristotle and the Theology of the Living Immortals*. Albany: State University of New York Press.

Broadie, Sarah. 2012. "The Ancient Greeks." in *The Oxford Handbook of Causation*, edited by Beebee, Helen, Christopher Hitchcock, and Peter Menzies. Oxford: Oxford University Press.

Chondros, Thomas. 2004. Dues-Ex-Machina: Reconstruction and Dynamics. In, *International Symposium on History of Machines and Mechanisms*. Proceedings HMM2004. Edited by Marco Ceccarelli, 87-104. New York: Kluwer Academic Publishers.

Dimarogonas, Andrew. 1992. "Mechanisms of the Ancient Greek Theater." Proceedings of the ASME 1992 Design Technical Conferences. 22nd Biennial Mechanisms Conference: Mechanism Design and Synthesis, 229-234. Scottsdale: ASME.

Endsjø, Dag. 2009. *Greek Resurrection Beliefs and the Success of Christianity*. New York: Palgrave Macmillan.

Hall, Edith. 2010. *Greek Tragedy: Suffering Under the Sun*. New York: Oxford University Press.

Johnston, Paul. 2019. "Rethinking Medea and the Deus Ex Machina." *Classical Philology* 114 (1): 125-135.

Merriam-Webster. 2022. Deus ex Machina. Accessed on January 11, 2022. https://www.merriam-webster.com/dictionary/deus%20ex%20machina

Merriam-Webster. 2022(b). Semantic. Accessed on January 11, 2022. https://www.merriam-webster.com/dictionary/semantic

Michelini, Ann Norris. 1987. *Euripides and the Tragic Tradition*. Madison: University of Wisconsin Press.

Milroy, James. 1992. *Linguistic Variation and Change: On the Historical Sociolinguistics of English*. Oxford: Blackwell.

Papadogiannis, Argyris S., Marilena C. Tsakoumaki, and Thomas G. Chondros. 2010. "Deus-Ex-Machina Mechanism Reconstruction in the Theater of Phlius, Corinthia." *Journal of Mechanical Design* 132 (1): 011001.

Roselli, David Kawalko. 2011. *Theater of the People: Spectators and Society in Ancient Athens*. Austin: University of Texas Press.

Shipley, Joseph. 1953. *Dictionary of World Literature Criticism Forms Technique*. New York: Philosophical Library.

Sourvinou-Inwood, Christiane. 2003. *Tragedy and Athenian Religion.* Lanham: Lexington Books.

Storey, Ian C. and Ariene Allan. 2014. *A Guide to Ancient Greek Drama. Vol. 4*. Hoboken: John Wiley & Sons.

Traugott, Elizabeth Closs, and Richard B. Dasher. 2001. *Regularity in Semantic Change. Vol. 97*. Cambridge: Cambridge University Press.

Traugott, Elizabeth Closs. 2014. "Semantic change." In *The First Glot International State-of-the-Article Book*, 385-406. Berlin: De Gruyter Mouton.

Watson, Peter. 2005. *Ideas: A history of Thought and Invention, from Fire to Freud*. London: Harper Perennial.

Worthington, Ian. 1990. "The Ending of Euripides' "Medea'." *Hermes* 118 (4): 502-505.

Wright, Mathew. 2008. *Eurpides: Orestes*. London: Bloomsbury.

Zuntz, Gunther. 1955. The Political Plays of Euripides. Manchester: Manchester University Press

CHAPTER FOUR

MECHANISMS IN THE MIDDLE AGES

A wooden windmill spins as steam raises from an ice-cold river on a brisk morning in a small agricultural settlement. Surrounded by lush black forests, citizens of a hilly village filled with single story shelters are up early in the morning plowing land and feeding pigs and livestock. Smoke pours out of chimneys as the sound of chimes and bells marks another day of labor–labor that necessitated a regular means of existence for the greater majority of people in Europe during the "dark" and subsequent Middle Ages. When the Roman Empire went into decline and eventual collapse, polytheism was all but eradicated across Europe. Major socio-cultural and political changes occurred which reflected the emergence of what many have since described as the Middle Ages–a time period of history that ranges from the 5th to 15th centuries. Throughout the Middle Ages, a number of developments arose including the establishment of some of the first parliaments, commerce, banking, institutionalized higher education, literary production, and arguably most important of all, processes which led to the abandonment of monarchs' divine right to rule.

Preceding these changes, the idea, usage and symbolic interpretation of the mechanism made its way into Rome. In theater, crowds continued to witness the mechane and would react with astonishment and ponder over the actions of gods (Cunningham 1954). Roman theater was alike to Greek theater and the usage of mechanes on stage remained prominent. In addition, the Romans used a mechane in funeral settings. The mechane went from being bound to forces that connected the populace with heterogeneous deities to eventually only being illustrative of a single deity. This occurred during rather turbulent and dictatorial times in Rome, and the shift was gradually solidified in the centuries after Caesar's death. On March 15, 44 BC, in the proximity of the Theater of Pompey, a group of Roman senators (with aid of upwards of 30 different conspirators) assassinated Julius Caesar. During Caesar's funeral, Antonius performed an encomium (a praiseworthy speech) in which he commended Caesar as a deity (Sumi 2015, 106). The funeral featured a wax manikin of Caesar that was propelled by a mechane into the sky in order to depict the God-like stature of the Roman. The wax figure reportedly contained twenty-three wounds over its body and face–portraying the actions that led to Caesar's death by stabbing.

While Roman funerals were indeed theatrical and ceremonial (to even greater extent than Greek), Sumi (2015) describes the usage of the mechane in Caesar's funeral as extraordinary (Sumi 2015, 108). To accompany the manikin, a mime was also put on stage to mimic Caesar. The mime named Caesar's murderers and voiced provocative and tragic lines (Sumi, 2015, 109). Afterwards, a major dispute ensued over the location of Caesar's proposed burial. On one hand, plebs attempted to take the body on their shoulders to the Capitolium for cremation, while on the other hand, various priests forcefully prevented this from happening (Sumi 2015, 110). This dispute was settled when the Senate provided funds from public treasuries to have Caesar cremated then buried at Velia (near Palatine Hill in Rome). Plebs continued to protest after the burial, meanwhile Mark Antony went onto marry Cleopatra (Caesar's intimate partner) which led to the subsequent break out of civil war between factions representing Antony and Cleopatra against Gaius Octavian's supporters. Octavian was triumphant and he then became the first formal emperor of the Roman Empire (he was also renamed as Caesar Augustus), while at the same time, he took on the role of a live deity. In terms of Augustus, by 12 AD he was proclaimed as the Pontifex Maximus–the dominant priest of all of the empire as well as the head of the Collegium Pontificum. This trend had begun with Plebs proclaiming Caesar's death as being a direct representation of God.

The impact of this shift in belief systems cannot be understated. The Catholic church not only emerged as an institution that sought to converge antecedent polytheistic spiritual belief systems, but it integrated itself with the polity, making the emperor of the Holy Roman Empire an all-powerful deity. This set the stage for a new objective reality for much of the world–a reality that Berdyaev refers to as the "Realm of Caesar." At this point in time, technological advancement was limited and societal assumptions about nature still reflected hierarchical order established by a theistic power (Berdyaev 1952, 48). Hence, from the height of the Roman empire to its decline and then into the Middle Ages, the mechanism remained theistic in ideation but was importantly marked by attributions to a specific single deity, rather than a heterogeneous collection of deities as in Ancient Greece.

Furthermore, in Rome, several centuries of relative peace (known as the Pax Romana) ensued under the sovereign Augustus and rulers after him, while religious conflicts occurred as pagans clashed with Christians until eventually (sometime in the 4th or 5th century AD) Christianity reigned paramount over paganism. Throughout the early segment of this intricate time period, a mass expansion of the Empire occurred under emperor Trajan (98-117 AD) and interestingly enough, an early Christian bishop from Syria described the role of the mechane with reference to the tenth book of the New Testament (Ephesians). Ignatius of Antioch was a disciple of John the Apostle and his articulation of the mechane can be traced in letters and teachings about Jesus Christ throughout the empire and Middle East. As many know, biblical scriptures state that Jesus Christ was crucified on a Cross. The Cross for Ignatius of Antioch is a cosmic hoist and mechane that propels people to heaven. He wrote, "O Cross, thou hoist to heaven! The Cross was driven into the ground–and behold, idol worship was destroyed. No ordinary wood is this, but the wood that God used for victory" (Ratzinger 2016).

Ignatius' articulation entails that the mechane does not cast down God into the masses but that the Cross raises people from their lowest level to Jesus Christ and in turn, propels them upwards into the heavens, closer to the realm of spirituality which is reigned over by God himself. In this sense, the emergence of a new historical idea can be observed about how common people (who were devout believers in Christ) could be raised out of paganism and into spiritual connection with one deity–God (at one point, the term Pagan literally meant non-urban peoples of the empire). Another noteworthy dynamic in Ignatius' articulation can be observed in how the Cross of Jesus Christ is the figurative mechane, whereas the rope to the mechane is the Holy Spirit (Ballard and Holmes 2006, 9). As such, the association between the mechanism and a theistic power was prevalent throughout the existence of the Roman empire and remained partially intact after its decline and eventual downfall. This conception then was carried over into the Middle Ages and reached a point of prodigious value in the late period of this era as observed in outstanding works of literature, a subject we now turn to.

The decline and eventual fall of the Roman empire begun in the early 5th century where Visgoths overthrew Rome, a city of one million inhabitants. These invasions led to a massive drop in population (Watson 2005, 243). An interesting, albeit understudied dynamic can be observed to have arisen over the course of several centuries that proceeded the collapse of the Roman empire. As noted by Watson (2005), even though Rome had the world's greatest libraries and collections of books, the Catholic Romans ardently rejected scientific inquiry and did not sponsor science in any format. The "banned" list of books and materials that the Vatican formulated and closely monitored disabled aristocratic circles of society to sponsor science and carry out scientific practices. The origin of the term censorship (censor) can be traced to Latin and particular practices of upholding and regulating societal morality in the Roman empire (Weaver 2019). With the decline of Rome came the immediate egress of the dark ages, a pair of centuries that have been widely discussed in popular culture as lacking any form of progress. Part of the reason behind such a dominant interpretation of history in the Western world is that Europe, at this time, lagged far behind China which was the most sophisticated intellectual civilization and was also the most technologically advanced (Watson 2005, 297). Chinese civilization already possessed paper, gun powder, coal mining, iron and steel manufacturing power, and widely reaching trade networks that spanned across the massive silk road. Meanwhile, in Western Europe, military invasions and economic decline were significant in these centuries, which along with cultural shifts and technological developments, contributed to a loss of texts (Rohmann 2016).

Two different developments arose from the sixth century to the middle of the ninth, argues Watson (2005, 249). These shifts fundamentally altered the way that ideas were exchanged, preserved, and communicated throughout societies. The arrival paper provided an alternative to papyrus which was used to write scrolls. A number of other technological developments arose over the course of several centuries in Europe including the watermill, the plough, crop rotation systems, the horseshoe, the blast furnace, and the clock (Watson 2005, 326). By the 12th century, universities started getting established and the ongoing development of the experimental method gained institutional support. As noted by Grant (2001) in *God and Reason in the Middle Ages*, the Middle Ages fostered various institutions in the west in commercial, educational, and religious spheres that went on to shape the modern world. Not only did such institutions have a lasting effect that still can be observed across social organization in our highly digital and technological societies of today, but these institutions enabled human beings to organize and govern themselves "that had never been done before" (Grant 2001, 1). Along these lines, the Middle Ages were shaped by developments that occurred largely throughout Western Europe. Grant argues that the arrival of a new self-consciousness based around reason combined with new educational activities (the establishment of universities) around 1200 led to the formation of intellectual life in Western Europe. These new processes were most visible throughout medieval universities. Indeed, while earlier civilizations such as the Mesopotamians or the Byzantine Empire did exercise forms of reasoning, a distinctive set of developments arose in medieval intellectual life within universities which turned out to be transformative as "reason in the university context was not intended for the acquisition of power over others, or to improve the material well-being of the general populace. Its primary purpose was to elucidate the natural and supernatural worlds" (Grant 2001, 3).

In all of previous human history, argues Grant (2001), no society had utilized reason in the way that those in the aforementioned university contexts ended up doing so. Needham (2013) contends that the decline of feudalism in Europe led to the rise of a new mercantile class that propelled the rise of science. Along these lines, prior to this period of history, reason did not encompass a central role that involved as many people over such an extended period of time. Grant's argument is crucial to reflect over because it echoes many other viewpoints of similar type. For example, Murray (1978) makes a similar set of claims. Starting around the 11th century, reason was used in the spheres of technology, magic, and astrology to deal with and attempt to overcome natural environmental forces (Murray 1978, 111-112). Upper echelons of society, including intellectual, economic, and political elites, all benefited from newly found usages of reason which concurrently served as a pathway towards upward mobility. Major changes and advancements to understandings of causality can also be observed in this time period of history. Most accounts on medieval understandings of causality begin with Thomas Aquinas and the time period of the 13th century. Aquinas made a distinction between essential and accidental causation and argued that there exist efficient causes and effects which we perceive and discover attuned to our senses. Each cause is prior to its effect entailing that "none can be its own cause, because it would then be prior to itself, which is impossible" (Marenbon 2012, 41). Hence, without a first cause, argued Aquinas, there would not be other causes and effects. The first cause was God and this cause was a "necessary First Cause" that was assumed to be ordered in a causal chain. Marenbon (2012) traces Aquinas' ideas to the metaphysics of Avicenna, a figure who also believed that God was the cause of existence (Marenborn 2012, 43). He also argues that Aquinas' contributions were seminal to an era that saw fundamental alterations in how the idea of the existence of God could be reconciled with contemporaneous philosophical frameworks of his time.

What ensued after Aquinas was a fundamentally different set of intellectual trends when compared to the previous 500 years. Since Aquinas posited that there exists some form of a natural underlying order of phenomena, this denied God's power of miraculous intervention (Watson 2005, 330). In turn, this led to a strand of thinking based upon "a totally new way of looking at the world" (Watson 2005, 330). Ultimately, the rise of the scientific method, precise measurement, and an intellectually unified secular world arose in Western Europe which formed the basis of Western modernity, argues Watson. Similarly, Grant (2001) makes a plausible claim in asserting that the consistency with which reason emerged in the 11th century "as a force to be reckoned with" (Grant 2001, 51), was due to a shift in the ideas of theologians in Western European spheres. This shift can be traced to how they dealt with questions pertaining to mysteries of faith. Evidence thus became more important than authority in theology (Grant 2001, 51). A large part of this transformative trend is attuned to Aristotle by Grant, as Aristotle's ideas and metaphysics constituted a primary source of rationality in Western thought. More specifically, Aristotle's use of reason represented a set of heuristics that were later drawn on by readers in Latin-speaking contexts throughout the Middle Ages as observable in their approaches to dealing with grandiose questions. Philosophers and theologians of the Middle Ages utilized an Aristotelian approach, as described by Grant, "In studying Aristotle, they could directly witness the way in which Aristotle employed reason to construct principles and generalizations about the world based on a modicum of observation that was often highly selective in order to justify this or that generalization" (Grant 2001, 97).

A variety of notions about the nature of causality arose during the early Middle Ages which can be observed throughout the works of Latin, Arabic, and Jewish philosophers (Marenbon 2012). The Muslim philosopher Ghazali's framework proposed that causal powers of different phenomena and forces are more determined by their natures than by theology. Similar ideas were put forward by John Scottus Eriugena, William of Ockham, Nicholas Autrecourt, Bernard of Arrezo, among others–all of which articulated philosophical frameworks through reading notions of causes and effects via a Neoplatonic lens (Marenbon 2012, 51). Similar to Aquinas, a Persian philosopher, Avicenna, articulated doctrines on metaphysical causation that posited causality is passed down via a causal chain from a theistic power (God) which was assumed to be "a necessary existent" (Marenbon 2012). In an assessment of medieval frameworks of causality, Marenbon comes to the conclusion that most frameworks on causation emerged out of these aforementioned traditions and tended to conceive of causes as things which led to modern readers pondering over the "ontological status of causes and effects, and to wonder at why the medieval authors left it so vague" (Marenbon 2012, 52). Causality in medievalist philosophical traditions was also marked by the denial of causal efficacy or necessity in nature.

Along similar lines, when it comes to the early moderns, Clatterbaugh (2012) argues that many philosophers attempted to tidy up the loose configuration of causes that were attributed by medieval thinkers to different material and immaterial phenomena and ideas about such phenomena. Ultimately, the early moderns did produce and confront a variety of different types of causal explanations, but still included both material and immaterial entities as causes in scientific explanations (Clatterbaugh 2012, 55). These processes, as the subsequent chapter of this book will demonstrate, were brought to an end when naturalist conceptions of causation came to the forefront alongside thinkers such as Newton and Hobbes. Ultimately, even though many of the noted thinkers drew from Aristotle, in the Middle Ages, interpretations of causation were fundamentally different when compared to Ancient Greece. Semantic changes occurred in the Middle Ages as the antecedent configuration of pre-Christian polytheism shifted into the formation of Christianity. This shift from a pagan-dominated societal worldview across Greco-Roman social orders had a profound impact on the subsequent development of reason, the enlightenment, and eventual hegemony of rationalism. Above all, it led to societal changes in how mechanistic forces were perceived–no longer were these forces of a heterogeneous theistic character as a single theistic force (Christ) took precedence.

The Prevalence of Christianity

The Middle Ages feature many of the characteristics that Berdyaev (1952) identified in his second period on human beings' historical relation to nature and the cosmos. The second period is marked by the institutionalization of ideas and messages put forward by Christ–these ideas shifted theistic beliefs in the West–albeit slowly from the time of Christ's crucifixion until the total decline of paganism by the end of the 4th century. When paganism declined it was gradually replaced by Christianity in nearly the whole of the European continent. Meanwhile, elementary forms of economics and serfdom also arose. Berdyaev argues that at this point, human beings did not yet overcome nature whether in elemental or biological form, but they nevertheless were under the prism of objectified religion. We must not downplay or fail to consider the immensely engrossing role that Christianity had in the Middle Ages on the lives of average Europeans. The church (whether Orthodox, Catholic, or Protestant), was incredibly dominant in its societal function and most commonly was aligned either directly or symbolically with the head of state (monarchical governance structures). In Western Europe, the majority of society was formed by peasantry who emerged alongside agricultural settlements and estates. From the sixth to tenth centuries, slavery (remnant from the Roman empire) throughout Europe progressively disappeared (Geary 2001, 187). Likewise, the distinction between slaves and free peoples became blurred.

From the late 9th century onward, the basic component of peasant society came into formation–the nuclear family unit (Geary 2001, 187). From this point on, populational growth ensued as did the foundation of many religious institutions which on a local level were mainly churches. Across societies, Christianity spread alongside the setting up of both monasteries and churches. Bishops and clergy became dominant in their societal roles and spread messages on the "basics of Christian belief" (Geary 2001). This included things such as acknowledgment of the trinity (Father, Son, and Holy Spirit), knowledge on heaven and hell, as well as the encouraging of honorific adoration of saints. For example, "for help in times of distress, whether periods of famine or illness, personal problems such as debt or conflict, loss of animals or oppression by local magnates, peasants made vows to their saints, which they then had to fulfill" (Geary 2001, 201).

In the system that grew to be dominant from 1000-1500, a parish structure took on a prominent role. This structure featured parishes and subdivisions–each of which had its own church and priest. This contributed to what Geary (2007) describes as a basic unit of Christian geography. Parish churches formed universal norms and dictated local traditions. They hosted baptisms, marriages, and burials. In the upper stratosphere of societies were lay and secular rulers that enjoyed various privileges which included appointing clerics to ecclesiastical office positions (Kennedy 2011). These types of collaborations, argues Kennedy (2011), enabled Christianity to spread throughout Europe at a rapid pace in the Middle Ages. In these expansive processes, churches were built and the aforementioned structures were instilled. The values of local people that had pre-Christian pagan rituals were converted into medieval conceptions of Christianity. Mass was widely attended and in conjunction with baptism, marriage, and burials, the church literally oversaw how the greater majority of populations in the Middle Ages experienced the most important events and trajectories in their lives. Similar prayers (such as

Hail Mary, Creed, and the Lords' Prayer), were memorized across the board which led to peasants understanding the "basics of Christian teaching" through incorporating them into their daily lives (Geary 2001).

The wide-reaching impact that Christianity had across societies in the Middle Ages has to be considered in relation to the bold fact that the peasantry and lower social orders were far different in terms of their access to information and knowledge than were the clerics (Hall 1990). The clergy spoke Latin and were representative of the group that would eventually break away into the enlightenment.

The Function of the Mechanism in Classic Literature

By the 14th century, many of the aforementioned inventions and ideas had taken hold in the European continent and the usage of the mechane remained prevalent. Italian architects improved countless aspects of theater engineering. For example, the Paradiso was a unique system of ropes hoisted by a machine that would elevate angels and other characters above stage (Brockett and Hildy 2007). Even with technological advances in areas such as theater and in urban life more broadly, populations still conceived and interpreted phenomena via a religious basis in daily life. Humans were still bound to mother-earth and were living in a tellurgic epoch (Berdyaev 1952, 48). This brings us to a crucial set of changes which impacted how social causality was constructed and interpreted. By the time the era of the enlightenment came around, novelists, play writers, and other literary contributors already started to shift away from conceptualizing the mechanism as a God-resembling force to simply incorporating external plot-changing forces that were not necessarily theistic. Examples include sorcerers, witches, natural climatic events, animals, among others. The most common manifestation attached to the deus ex machina, argues Ray (2008), is in powerful figures such as that of a King (Ray 2008, 138). In words, the heterogeneity that was inherent to the Ancient Greeks' conception of the mechane re-emerged, but it was absent of necessary theistic attachments to causality. Alongside this newly emergent variation in the deus ex machina, the overarching role of religion in daily life in Europe should not be downplayed. Hans Christian Andersen's writings and short stories are but just one example of how the Christian conception of God's causal role in life was interpreted by populations leading up until the 19th century. For example, in the short story, *The Old Church Bell*, Andersen (1983) wrote the following remarkable introduction,

> "It was late in autumn. The leaves still clung to the grape-vine, but they were already tinged with red. Rainy gusts swept over the country, and the cold autumn winds increased in violence and roughness. It was no pleasant time for poor folk. The days became shorter and gloomier; and if it was dark out in the open air, in the little old-fashioned houses it was darker still. One of these houses was built with its gable end towards the street, and stood there, with its small narrow windows, humble and poor enough in appearance; the family was poor, too, that inhabited the little house, but good and industrious and rich in a treasure of piety concealed in the depth of heart. And they expected that God would soon give them another child; the hour had come, and the mother lay in pain and sorrow. Then from the church tower opposite the deep rich sound of the bell came to her. It was a solemn hour, and the song of the bell filled the heart of the praying woman with trustfulness and faith; the thoughts of her inmost heart soared upward towards the Almighty, and in the same hour she gave birth to a son" (Andersen 1983, 6).

Hundreds (if not thousands) of other tales, short stories, and plays from the Middle Ages feature similar theistic logics that determined social outcomes. With that being said, there were also a number of prominent non-theistic implementations of the deus ex machina which makes the Middle Ages heterogeneous compared to that of Ancient Greece. Below, three different examples will shed light on the usage and evolution of the deus ex machina.

The Tale of Gamelyn

The 14th century English tale Gamelyn (1350-70) features a noteworthy manifestation of a plot-shifting mechanism in what can be considered to be a non-traditional romantic tale. Although there is no directly verifiable evidence of Gamelyn's origin (Edwards 2011), historians have noted that it is a part of a collection of tales from medieval literature referred to as the "Matter of England," and was originally included as part of Geoffrey Chaucer's manuscripts in the *Canterbury Tales*. The latter are comprised of 29 different manuscripts including the notoriously uncompleted, *Cook's Tale* (Edwards 2011, 79). Gamelyn begins in the context of the regime of King Edward I (the mid-13th century)–a time that was marked by war, conquest, assassinations, and rebellions. When on his death bed, Sir John of Boundis lay faced with a decision to divide his wealth between three of his sons–Ote, Johan, and Gamelyn. Right after his father's passing, the oldest son, Johan, ended up unjustly taking all of the share (Sands 1986). At that time, the main character, Gamelyn, was still a young child. Even though he was a child, Johan mistreated him and also took over his land inheritance. After spending much of his life as a servant to his brother and after the death of their father, Gamelyn finds that obtaining a property inheritance is overly difficult due to corruption in law and his brother's authoritative will.

Gamelyn and the accomplices that join him throughout his attempt(s) to obtain the inheritance have been described by literary historians as being emblematic of a common or ordinary people who have "enduring values" on their side (Knight and Ohlgren 1997). Along with his accomplices, Gamelyn storms into Johan's castle and breaks down the doors. The castle is overtaken and Gamelyn hosts a large feast. Meanwhile, Johan is attempting to get his property back and is concurrently plotting revenge against his brother. After Gamelyn's guests leave the castle, Johan calls authorities who then capture Gamelyn and imprison him. The reader is faced with no way to imagine how the protagonist can get out of this conundrum, only to have the middle brother (named Sir Ote) come and rescue him. This action represents the usage of a deus ex machina in the story (Johnston 2010, 54). Subsequently, Gamelyn is free and his accomplices assault the judge and takes his place in the courthouse. His accomplices then take the place of the jury and order Johan as well as the corrupt authorities to be hanged. Ote and Gamelyn then go to the King who grants Ote a position as justice and Gamelyn is made Chief Justice of the forest (Sands 1986). Both Ote and Gamelyn then split and share the inheritance and the story ends on a happy note.

It is evident that this mechanistic plot solution in this story is not theistic in its character, which illustrates how provides a prominent example of writers still relied upon the mechanism as an articulation of a forceful action and event, but did not link it to theistic powers. Problems in a given plot were solved via the mechanism and when this transpired, the reader or viewer would be forced into a major dilemma as the structure, cohesiveness and transcendent worth of a given plot would either fall into question or be resolved. The mechanism remained influential into the great classics of literature and theater.

It is evident that this mechanistic plot solution in this story is not theistic in its character, which illustrates how provides a prominent example of writers still relied upon the mechanism as an articulation of a forceful action and event, but did not link it to theistic powers. Problems in a given plot were solved via the mechanism and when this transpired, the reader or viewer would be forced into a major dilemma as the structure, cohesiveness and transcendent worth of a given plot would either fall into question or be resolved. The mechanism remained influential into the great classics of literature and theater.

Shakespeare's *As You Like It*

Perhaps the greatest writer in the history of the Western world, William Shakespeare's works have defined entire genres and influenced hundreds of millions (if not billions) of people throughout the last 500 years. One of the better-known implementations of the deus ex machina by Shakespeare can be observed in his comedy, *As You Like It* (written in 1599). Likewise, Shakespeare has been noted to utilize the deus ex machina more frequently in his comedies when compared to other plays and stories. In contrast to the previous examples provided in this book, Shakespeare utilizes the deus ex machina three different times in one single story–a story about a woman named Rosalind who along with her cousin Celia, flee legal persecution and flee to the forest of Arden in search of Rosalind's lover, Orlando. The meaning of this comedy has been interpreted as representing an ideal of a Renaissance Christian ethic (Cirillo 1971).

The story opens up with Sir Rowland de Bois on his death bed. While taking his last breath, he instructs his son Oliver to care for his younger brother Orlando. What ensues is domination and abuse. The older brother repudiates his younger brother's access to different facets of life and resources. Meanwhile, a Duke named Senior has been expelled from his throne by his own brother, Duke Frederick. Senior then flees to the same forest and lives a lawless life with a group of accomplices. Senior has a daughter named Rosalind who is close friends with the daughter of Frederick, Celia. While Frederick is fleeing to the forest, Orlando gets in a fight with a court wrestler named Charles and loses. After the loss, Orlando falls in love with Rosalind.

In the development of their relationship, Rosalind hides her love of Orlando from Celia. Frederick then banishes both Orlando and Rosalind. Rosalind goes into the forest with Celia. Whilst in the forest, as a disguise, Rosalind dresses as a man and called herself Ganymede and Celia dresses up as a shepherdess named Aliena. Meanwhile, the Duke, Frederick, is very angry at his daughter's actions and connects Orlando's disappearance with Rosalind's. This brings us to the first manifestation of Shakespeare's usage of the deus ex machina in this story: when in the forest, Oliver is preyed upon by wild beasts (a snake and lion) while sleeping and even though Oliver treated his own brother poorly, it is Orlando who rescues him from these beasts. This results in Oliver being converted "from a villain to innocent within the confines of Arden after he had been saved from both the serpent and lioness by Orlando in a Christian act of turning the other cheek" (Cirillo 1971, 33). Along these lines, Cirillo (1971) argues that the unnatural environment of the legal institution in the story (the court), is restored into a natural environment that is restored within the forest of Arden after the serpent and lioness are expelled.

Subsequently, Frederick sponsors Oliver to put together a search party and through these plans, the Duke goes directly after his brother Duke Senior and his lands. At this time, Senior is living in the forest with a set of roving bandits. Senior then comes across Orlando in the same forest and takes him under his care. Nearby, Rosalind and Celia obtain access to a cottage in which they live in. Rosalind then meets Orlando who does not identify her because of her disguise. Orlando thinks the character of Ganymede is real and agrees to a proposition to help Orlando deal with unhappiness due to missing his lover. In the course of these lessons, another woman, Phoebe, ends up falling in love with Ganymede. Interim, Oliver falls in love Aliena (Ciela)–both agree to get married. Subsequently, the wedding is organized by Ganymede and is attended by the following characters: Phoebe, Silvius, Oliver, Celia, Audreny, Touchstone, Orlando, and Hymen–all of which then go to Duke Senior and his followers. During the wedding, both the disguised Rosalind and Celia leave then quickly return as themselves, out of their disguise, alongside Hymen (the God of marriage)–who then carries out the marriage procession. The inclusion of Hymen into the story is considered to represent a deus ex machina because after his arrival Rosalind returns back to her normal self, realigns with her father, and becomes Orlando's wife. Likewise, as noted by Werkmeister (2015), "through marriage, time (which collapses in the forest) is restored–the names, titles, and land of the ancestors (i.e., the past) are propelled forward through the promise of children (i.e., the future)." Hymen impacts the lives of all characters who seem to be lost and tangled in the forest (Rosalind and her tie with Orlando; Celia and Oliver; Audrey and Touchstone), as his arrival enables all entanglements to be resolved.

Furthermore, while making statements conjoining the bride and groom, the wedding is interrupted by an army led by Duke Frederick. On his way to the wedding, Frederick encounters an elderly man who is highly devout and religious. He is portrayed as a holy figure that convinces Frederick to disregard his material desires and not show anguish towards his brother. Shakespeare describes this as a conversion that results in Frederick's retirement from the world and earthly things. This third and final usage of the deus ex machina is perhaps the most significant of all three in this particular story as the elderly man is not only a hermit, but is someone that is a part of the wilderness and the forest itself (Werkmeister 2015). In the end, through the author's implementation of the deus ex machina, it is the forest itself that functions as a place of refuge for characters and also as an area that is metaphysically capable of transforming and resolving the plot of the story. The forest of Arden is a circumscribed world and is a second world within the fiction story just as the fiction story is a second world with relation to the audience (Cirrillo 1971, 35).

Hamlet

Another prominent manifestation of a Shakespearean utilization of the deus ex machina can be observed in arguably his greatest and most well-known tragedy, Hamlet (1599-1601). Not only is this tragedy a cultural staple of Western theater, but it is also Shakespeare's longest work (nearly 30,000 words in length). The tragedy takes place in Denmark and is based around the main character, Prince Hamlet, who seeks revenge over his uncle Claudius who killed Hamlet's father in a power-grab over the throne (Draper and Draper 1966). A theme of this tragedy is in the main character's lack of ability to form a comprehensive set of views and course of actions to take in response to his family being undermined. The first act begins with a ghost of the King of Denmark informing his son Hamlet that he has to avenge his father's death. The setting of the first act is in the Elsinore castle–a large fortified square shaped structure with numerous towers surrounded by lush green grass, a moat, and a harbor that makes its way to the North Sea. Late in the evening, a friend of Hamlet's named Horatio arrives to the castle and tells Hamlet about the ghost as well as news of Claudius not only overtaking the throne, but also marrying his mother. Hamlet's clear disappointment with this news marks an overall theme of anguish in the first act that is combined with him questioning his mother's loyalty to his own family. The following day, a royal member named Polonius tells his son Laertes farewell before he departs for a trip to France and meanwhile, Laertes informs Ophelia, his sister, to not engage with Hamlet in a romantic manner.

The second act is marked by Hamlet's increasingly divergent behavior in which he attempts to abandon romantic relations with Ophelia. In the background, Claudius (along with his close assistant Polonius) are spying on Hamlet in order to rationalize this new change in behavior. These spying inquiries deepen until a group of actors arrives to the castle who know Hamlet. Hamlet then writes a story for the actors to enact in which the murder of his father forms the central plot content. Further, in the third act, Hamlet is observant of Claudius' response to the short play, and as expected, the latter reacts adversely by leaving the hall. He then sends Hamlet out of the castle's premises. Through his experiences outside of the castle, Hamlet deals with his mother and reasons that if he kills the King, then his soul will make a metaphysical journey into heaven rather than into hell. Although the decision to actually kill the King is not formalized in this act, the plans appear to be set. However, abruptly, the end of the act is marked by an actual murder–as Hamlet kills Polonius who during Hamlet's meeting with his mother (Queen Gertrude), is spying in the background. At this ending point of the act, Hamlet's father's ghost re-emerges and tells Hamlet to not take any further steps to pursue vengeance.

Into the fourth act, the King of Norway (Fortinbras) is set to cross Denmark with his army in an offensive against Poland. Hamlet goes to England by ship but surprisingly comes back. In a letter sent to Horatio by Hamlet, it is written that pirates attacked the ship and took Hamlet into their custody as a captive. They brought him back to the castle in exchange for a favor that Hamlet promised them–a favor that would see Hamlet deliver letters on the pirate's behalf to the king. Hamlet's return back to the castle is considered to be a dramatic return and one that constitutes usage of the deus ex machina. Theatrical enactments of this part of the tragedy during the Middle Ages featured the usage of a mechane device that would move Hamlet, via the pirates, back to a part of stage that was decorated as Denmark and the aforementioned castle. Hunt (2007) makes a compelling argument by stating that the usage of the deus ex machina here by Shakespeare turned out to be historically transformative because it is not a theistic power that directs Hamlet, but rather, reality becomes internalized (Hunt 2007,153). Hunt also argues that the ghost is a major part of this process because of its materialization multiple times in the first act (Hunt 2007, 154), with relation to its final appearance. The ghost's projection is eventually something that is produced within Hamlet's brain and is directed through the individual mind–an entity that Hunt (2007) describes as a "nutshell that contains infinity" (Hunt 2007, 154).

On his journey back Hamlet comes across evidence that Claudius has been plotting to kill him upon his arrival. He thus returns to Denmark and the castle and sends two associates, Guildenstern and Rosencrantz, to go to England in his place. The act ends with the drowning and death of Ophelia who takes her own life because of Hamlet's romantic rejection. In the final act, Hamlet encounters Horatio in a graveyard and both contemplate their future and the meaning of death. Meanwhile, Ophelia's brother, Laertes, shows up to the graveyard along with the funeral procession for his dead sister. As expected, Laertes feuds with Hamlet and an altercation breaks out which eventually turns into a declaration for a duel. Prior to the duel, Claudius forms a strategy with Laertes to kill Hamlet through the use of poison. A vase of wine is marked with poison and during the actual duel, the queen unknowingly drinks from a cup with the poisoned wine and quickly dies. Meanwhile, Hamlet as well as Laertes are cut by a blade which also has poison on it–the latter dies first (from a self-inflicted, accidental cut), while the former has enough time to stab and kill Claudius. Before dying, Hamlet forces Claudius to drink from the remaining poisoned wine which leads to his death. Upon return from armed conflict with Poland, the King Fortinbras returns alongside English statesmen who are met by this gruesome situation and an explanation from the only remaining person, Horatio.

Hamlet may in fact be the most significant work from the end of the Middle Ages that utilizes the deus ex machina in a way that sets the reader (and the socio-cultural configurations of the time) into a trajectory of self internalization and individualism. The theistic underpinning of the "god from the machine" starts to deviate away from causal forces of the cosmos and theistic powers. This is likely why Hegel considered Shakespeare's Hamlet as the "supreme modern play" (Hunt 2007), or as an archetypal modern drama that is marked by character and not by a collision of universal forces (Bates 2010, 19). On the other hand, Pascucci (2013) makes a fascinating set of claims in assessing Hamlet's relationship to the ghost. This relationship features both substance and shadow, both reason and madness and a material being that meets an incorporeal being. From this viewpoint, Pascucci (2013) contends that Hamlet can be viewed as a prophet because he possessed knowledge made of corporeal ideas as well as a material form of knowledge comprised of sensible ideas. Hamlet may embody an emerging set of ideas on humanism that were on the rise at the end of the Middle Ages.

Indeed, these stories, plays, and novels do not cover the vastness of literature and socio-cultural creations that were prevalent throughout the Middle Ages, but they do, nevertheless, serve as exemplars of how some of history's most coveted writers utilized the deus ex machina to resolve plot twists and demonstrate what at the time was a dominant societal assumption of causation being theistic or miraculous in its nature. The Middle Ages also turned out to be the last period in human history that was absent of large-scale machinery and the general mechanization over nature. Once the industrial and scientific revolutions emerged, characteristics of humanity's relationship to not only itself, but to the world and the cosmos fundamentally transformed–a transformation that set us on a path of no return, a path towards the solidification of nation states, the collapse of empires, the rise of international institutions, humanistic ideologies, the discovery of bacteria, modern medicine, vaccines, atoms, and a highly particular, positivist, and hegemonic set of assumptions on the nature of causation and mechanisms in which the phenomenal world became prominent, and the noumenal slowly declined into an abyss.

References

Andersen, Hans Christian. 1983. The Complete Illustrated Stories of Hans Christian Andersen. London: Chancellor Press.

Ballard, P., & Holmes, S. (2006). eds. The Bible in pastoral practice: Readings in the place and function of Scripture in the church. Wm. B. Eerdmans Publishing.

Bates, Jennifer Ann. 2010. Hegel and Shakespeare on Moral Imagination. Albany, NY: SUNY Press.

Bell, James. 2013. The Spiritual World of the Hobbit. Bloomington: Bethany House Publishers.

Berdyaev, N. 1952. *The Realm of Spirit and the Realm of Caesar*. Trans. by Donald A. Lowrie. London: Victor Gollancz.

Brockett, Oscar G., and Franklin J. Hildy. 2007. *History of the theatre: Foundation edition*. Boston: Allyn & Bacon.

Cirillo, Albert R. 1971. "As You Like It: Pastoralism Gone Awry." *ELH* 38 (1): 19-39.

Clatterbaugh, Kenneth. 2009. The Early Moderns. in *The Oxford Handbook of Causation*, edited by Beebee, Helen, Christopher Hitchcock, and Peter Menzies. Oxford: Oxford University Press.

Cunningham, Maurice. 1954. Medea ΑΠΟ ΜΗΧΑΝΗΣ. *Classical Philology* 49 (3): 151–160.

Draper, John, and John William Draper. 1966. *Hamlet of Shakespeare's Audience*. Hove: Psychology Press.

Edwards, Anthony, Stockwell Garfield. 2011. The Canterbury Tales and Gamelyn. In *Medieval Latin and Middle English Literature: Essays in Honour of Jill Mann*, edited by Mann, Jill, Christopher Cannon, and Maura Nolan. Suffolk: DS Brewer.

Geary, Patrick J. 2001. "Peasant Religion in Medieval Europe." *Cahiers d'Extrême-Asie* 12: 185-209.
Grant, Edward. 2001. *God and Reason in the Middle Ages*. Cambridge: Cambridge University Press.
Hall, David D. 1990. *Worlds of Wonder, Days of Judgment: Popular Religious Belief in Early New England*. Cambridge: Harvard University Press.
Hunt, Marvin W. 2007. *Looking for Hamlet*. New York: St. Martin's Press.
Johnston, Andrew James. 2011. "Wrestling in the Moonlight: The Politics of Masculinity in the Middle English Popular Romance Gamelyn." In *Constructions of Masculinity in British Literature from the Middle Ages to the Present*, 51-67. New York: Palgrave Macmillan.
Kennedy, Philip. 2011. *Christianity: An Introduction*. London: IB Tauris.
Knight, Stephen and Ohlgren, Thomas. 1997. *Robin Hood and Other Outlaw Tales*. Kalamazoo: Medieval institute Publications.
Lynch, Stephen. 2003. As you Like it: A Guide to the Play. Westport: Greenwood Press.
Marenborn, John. 2012. The Medievals. The Oxford Handbook of Causation. New York, NY: Oxford University Press.
Murray, Alexander. 1978. *Reason and Society in the Middle Ages*. Oxford: Oxford University Press
Needham, Joseph. The Grand Titration: Science and Society in East and West. Vol. 21. Routledge, 2013.
Pascucci, Margherita. 2013. *Philosophical Readings of Shakespeare*. New York: Springer.
Ratzinger, Joseph Cardinal. 2016. *Joseph Ratzinger Collected Works: Theology of the Liturgy*. San Francisco: Ignatius Press.
Ray, Ratri. 2008. *William Shakespeare's As You Like It*. New Delhi: Atlantic publishers & Distributors.
Rohmann, Dirk. 2016. *Christianity, Book-burning and Censorship in Late Antiquity: Studies in Text Transmission*. Berlin: De Gruyter.
Sands, Donald. 1986. *Middle English Verse Romances*. Exeter: University of Exeter Press.
Shakespeare, William. 2000. As You Like It. In *The Pelican Shakespeare*, edited by Ed. Frances E. Dolan. New York: Penguin.
Sumi, Geoffrey. 2015. *Ceremony and Power: Performing Politics in Rome between Republic and Empire*. Ann Arbor: University of Michigan Press.
Watson, Peter. 2005. *Ideas: A history of Thought and Invention, from Fire to Freud*. London: Harper Perennial.

Werkmeister, Steve. 2015. “Shakespeare’s As You Like It.” Accessed on January 5, 2022. https://stevesofgrass.wordpress.com/2015/07/24/shakespeares-as-you-like-it/

Weaver, Kari D. 2019. "Censorship in the Digital Age the World Over." In *Advanced Methodologies and Technologies in Government and Society*, edited by Mehdi Khosrow-Pour. 481-492. Hershey: IGI Global.

CHAPTER FIVE

MECHANISMS IN THE NATURAL AND SOCIAL SCIENCES

Wooden tables and shelves full of open books, glass containers, rulers, and small hoists fill a Victorian room with high ceilings. Containers with chemicals are moved around by scientists using scalpers to attempt to identify properties of chemical reactions. Outside, carriages pass by and hundreds of people wearing tall top hats go about their busy day. Such settings symbolize an era that was marked by the emergence of the scientific revolution and a rise in experimentation. Throughout urban areas, an age of invention and humanistic progress arose after the enlightenment (which many consider to date to the 15th century). No single factor can be considered to have brought about the enlightenment, and just as with many other complex social developments, a configuration of factors likely contributed to its emergence. As Watson (2005) reasons, the advent of the clock improved measurement and planning, the importing of paper producing capabilities and writing from China expanded literary capabilities, and the impact of the great plague of the 14th century resulted in people shifting towards more inward forms of faith and away from collective religious circles. All of these developments and events had their own impact on the enlightenment era's emergence (Watson 2005, 389). Likewise, the rise of trade and seafaring in Italian city states (e.g., Venice; Genoa; Florence) led to an explosion in commercial practice and the establishment of the first modern banking systems. As noted by Watson (2005), "for individuals, aware that they had to rely on their own strengths, conscious of the superiority of rationality over tradition, and that time and money-management were the key, life took on a faster pace." (Watson 2005, 391). This is when humanists first emerged and changed the curriculum of human knowledge (Watson 2005, 391).

Humanism provided socio-cultural and political alternatives to the divine order that marked Christianity and turned populations away from a highly religious way of life. This fostered many new ways of viewing the world, argues Watson (2005), and in turn, likely contributed to the enlightenment's profound effect. This shifted the pursuit of virtue into individualistic avenues. In the context of Britain, huge changes occurred as monasteries, chantries, and religious figures were sold off and nobles were replaced by commerce and craftsmen (Watson 2005, 419). The early 16th century was marked by the peak of the renaissance and figures such as Leonardo Da Vinci or Michelangelo made timeless contributions to human knowledge. Not long after, in the North of what is today Germany, Martin Luther spurred what historians frequently refer to as an "accidental revolution" (Hollingsworth 2013, 17). Luther's ideas led to the formation of Protestantism (and later, Calvinism), which rejected the authority of the Catholic church and led to a split in Western Christianity. Whereas Erasmus had sought to free Christians from what many would describe as antecedent superstitions, he did not intend to "interfere with Christianity's basic theological framework," Luther's ideas were momentously transformative because they enabled segments of Christian populaces to engage in their faith alone and interpret scripture alone. Hence, some started believing that religion could be practiced in the word alone (Ryrie 2018). Ryrie (2018) argues that in the time period of Luther's emergence, the new technology of printing ended up finding its "first master." The mass-market pamphlet then became a quick and cheap method with which to spread information across large scales of people.

By the mid-16th century, views about the cosmos started to become more scientific. If Luther's contributions to the formation of Protestantism spurred what historians refer to as an accidental revolution, Copernicus in his *On the Revolutions of the Celestial Spheres* (1543), triggered the scientific revolution. Copernicus identified that the sun was the center of the solar system rather than the earth. This brought about a huge contagion of criticisms as Copernicus' ideas challenged the previous status quo of human knowledge–no longer was the earth the center of the universe. Copernicus was a Polish cleric in the hierarchy of the Catholic Church, and as argued by Repcheck (2007), he became tainted by the brush of the Lutheran Reformation. The subsequent major work that is commonly mentioned after Copernicus is Galileo's *Dialogue Concerning the Two Chief World Systems* (1632), a work in which he compared Copernicus' framework to the Ptolemaic system. The social and cultural championing of Copernicus' heliocentrism led to Galileo's imprisonment. Authorities argued that Galileo challenged the socio-political status quo of the time. Interestingly enough, this occurred just before the onset of the inquisition. Furthermore, the next logical development in the historical trajectory of the scientific revolution came with Isaac Newton's *Principia Mathematica* (1687) and the identification of laws of motions, gravity, and a large number of other advancements in mathematics, including the solidification of calculus.

On the whole, the contributions of the many ensuant thinkers to the scientific revolution are too vast to go into detail, yet after Newton, the following names deserve mentioning: Vesalius; Halley; Harvey; Kepler; Descartes; Pascal; Franklin; Russel; Mendeleev; Dalton; Darwin; Mendel. Francis Bacon is widely recognized as being the founder of the scientific method. The scientific revolution transformed the basis of human knowledge and shifted society towards individualism, atheism, and empiricism. These innovative ways of thinking were premised on a new found sense of humanism. In 1625 in Germany, 45.8% of publishing activity in Germany was based on theological themes. By 1800, this number dwindled down to only 6% (Watson 2005, 509). Turning back to Berdyaev's (1952) aforementioned account of the historical relation that human beings have to cosmos, the scientific revolution occurred in the third of Berdyaev's identified periods. Berdyaev described how human beings carried out a process of mechanization over nature–a process that was reliant on two factors–those of scientific and technical control. Here, understandings of social causality changed greatly. No longer were causal forces in nature and daily life linked to theism, but rather, the rise of scientific inquiry enabled widespread cognition to emerge about the properties and materials that make up the phenomenal realm. Along these lines, the rapid development of industry resulted in stellar cognitive changes in how human beings went about understanding material reality. This was accompanied by a gradual decline in faith and the separating of the state from religious institutions. By the beginning of the 19th century, the myth of humanism was broken, argued Berdyaev. This led to an abyss opening beneath humanity. Berdyaev conceptualizes this abyss through linking it to the egress of capitalism and the marketization of economic life. A profound implication arose from marketization–for the first time in history, masses of people were necessitated to sell their labor for wages under specified industrial trades.

Increased industrialization, urbanization, and the decline of feudalism changed the structure of economics. The relationship that human beings had with society and politics also transformed. These processes entailed major changes for organized religion. The authority of leading churches, such as the Catholic or Orthodox Church, was eventually stripped away and transferred over to the state–this resulted in a newly found objective totality belonging to statehood. These dynamics, as one would expect, eventually led even greater adverse outcomes, specifically in the rise of the totalitarian state and totalitarian ideologies–both of which featured common recognition of "the complete authority of the state and of society" (Berdyaev 1952, 57). The third period exemplifies how the realm of Caesar always strives to subject itself to the whole of humanity in its universalizing tendency. This pattern has turned out to be the main tragedy of history, of freedom, of necessity, and of the human being's destiny (Berdyaev 1952, 79). This period of history resulted in the forging of a comprehensive study of the phenomenal world. In the English lexicon, such a historical development ended up being interpreted as constituting natural science which a noun and is defined as, "any of the sciences (such as physics, chemistry, or biology) that deal with matter, energy, and their interrelations and transformations or with objectively measurable phenomena" (Merriam-Webster, 2021a). In contrast, as subsequent sections of this chapter will reveal, social science differs in its nature and while also being a noun, social science is defined as, "a branch of science that deals with the institutions and functioning of human society and with the interpersonal relationships of individuals as members of society," or "a science (such as economics or political science) dealing with a particular phase or aspect of human society" (Merriam-Webster, 2021b). Before delving into these differences, attention must be given to the antecedent conditions that enabled an era of modernity to arise alongside science.

From Theism to Atheism

With the onset of modernity, populations in much of the Western world experienced a decline in religious beliefs. This was a slow-moving process that did not unfold in one year, ten years, or even one hundred years, yet it did pave way for what turned out to be a truly a radical change in human thought. The era of the enlightenment, the onset of modernity, and the emergence of liberalism and industrialization contributed to masses perceiving themselves as being free from God's oversight of their lives. No longer did swaths of populations live under the prism of the church in parish structures. Millions moved into urban areas and communal lifestyle gradually began to decline as greater property rights emerged and individuals were able to disconnect from non-urban areas of life. Alongside these processes, the separation between church and state as well as the general mechanization (human control) of nature started to be achieved in Western and Central Europe. With mechanization of nature, human beings, for the first time, developed a substantial degree of control over natural forces which has since only intensified and increased in scope over the last several centuries.

Likewise, with the emergence of modernity, nature ceased being interpreted as a natural hierarchical order controlled by theistic forces. Descartes was a principal figure here as his ideas contributed to objects being studied independently of the mind. During this era, science and the scientific method developed and in time, became paramount in their influence on society. Nietzsche's philosophy epitomized these changes, especially in his idea of Gott ist tot (the Christian God is dead). Scientifically, the solidification of the principle of reason (as articulated by Leibniz in the 17-18th centuries) was significant in its consequences of changing the overall course of previous millennia of human history. Leibniz's principle entails that everything in the world must have some naturalistic reason or cause behind its existence. For instance, if there is a given entity (z) then there has to be an explanation for why it exists. The same exact logic is applied to every given event and every given proposition. Such ideas contributed to a new trend–for the first time in history, natural phenomena could be explained and, in some cases, predicted by physical laws. For example, the law of the conservation of mass was discovered in the eighteenth century. This law states that in any closed system, a given mass of substances that are produced by a chemical reaction are always the same as the reactant. In other words, in any given chemical reaction, mass is unvarying and atoms do not disappear, say when say a piece of wood burns into ashes. Instead, atoms get reshuffled and reconfigured. As many other findings, this discovery helped change the status and position that science had over daily life and society.

Since nature was no longer dictated by theistic forces, understandings of causality and mechanistic forces changed greatly. No longer were causal forces in nature and daily life linked to theistic powers, but rather, the rise of scientific inquiry enabled widespread cognition to emerge about properties and material of the world. Mechanisms were assumed to be operative as forces in the phenomenal realm that could be discovered by human beings. The idea of mechanisms fundamentally shifted from entities that were uncontrollable or constructed to entities that could be measured and importantly, discovered in the empirical world. Along these lines, crucially, the interpretation of causality and causal forces began to lose ties to their age-old polytheistic and theistic characteristics.

A very fascinating comparative angle to these changes is provided by Watson (2005) who weighs the rationale that indigenous populations in the "new" world had in comparison to Western Europeans. In the new world, there was such a thing as science, argues Watson, but it was based on primitive technology and simple machines that the Europeans had used in contexts such as Ancient Greece. An interesting observation here is made by Watson (2005) in that some indigenous populations in the new world made and used simple machines based on classical Greek mechanics including the wedge, an inclined, plane, the lever, the pulley, and the screw (Watson 2005, 452). Native Americans used these devices in tree-felling and canoe-building which brings us to a bold contrast that set these populations apart from Europeans. As noted by Watson, "whereas Europeans by the fifteenth century were searching for ultimate causes, whose outcomes could always be predicted, native Americans preferred to control the forces of nature by means of intimate relationships with the spirits that controlled these forces–achieved through ritual or dreams (Watson 2005, 452). This clear distinction in ideational characteristics of different civilizations and peoples is crucial to acknowledge because it was during the onset of modernity and the end of the Middle Ages when Europeans started to believe that the natural world was ruled by laws.

Into the modern period of history, novelists and playwrights continued to utilize the deus ex machina, but the overall societal conception of causality had changed. Implementations of the deus ex machina tended to be absent of any polytheistic and theistic components, and as time went on, the inclusion of such plot twists or resolutions was simply done to resolve stories. For example, the deus ex machina can be observed in Tolkien's *The Lord of the Rings* and *The Hobbit*, in Dickens' *Oliver Twist*, in Golding's *Lord of the Flies*, in H.G. Wells' *The War of the Worlds*, among numerous other works that are considered to among the most important classics of modern literature. In Tolkien's works, a key character (Gandalf) is a sacred wizard that gets supplemented and rescued in near-death situations by eagles. These "Gwaihir" eagles as Tolkien refers to them, are the leaders of an elite set of eagles that soar over the high skies and operate as a plot altering mechanism to save the wizard in times of great danger. The eagles are portrayed as being large enough to carry and transport humans on their back. In Tolkien's novels, the eagles function as agents of rescue and wrath, argues Bell (2013). In *The Hobbit*, during the battle of the five armies, at a crucial point in the story's plot, the eagles arrive and their arrival tips the fight in favor of those on the side of the hobbits. In *The Fellowship of the Ring*, an eagle rescues Gandalf when he is trapped on top of Saruman's tower. Likewise, in *The Return of the King*, a number of eagles help protagonist Aragorn during a battle and also rescue main hobbit characters Sam and Frodo (Bell 2013).

Other utilizations of mechanisms as plot shifting or solving tools can be observed in noteworthy novels such as Golding's *Lord of the Flies*. When a large group of adolescent boys are stranded on an uninhabited tropical island, they form factions and end up abusing one another. At the most intense segment of the plot, a central character to the story, Ralph, is facing a near-death situation. Then, out of nowhere, Ralph gets rescued due to the appearance of a ship which metaphysically pulled him away from a violent charging crowd of his peers. The ship was a British naval vessel that embodied order and safety; it symbolized a contrast to the anarchistic conditions that arose on the island. With the onset of modernity and the subsequent scientific revolution, a fundamental alteration arose to the meaning of what constitutes mechanistic forces. This alteration led to a huge increase in the discursive usage of the term mechanism.

The Discursive Explosion of the Mechanism

With the rise of modern science, the etymological basis of the Greek *mechane* paved way for the English term, mechanism. This term automatically took on the following meaning(s): a part of a machine, a machine, a system of interacting parts, or an intermediate force that lies between cause and effect. As Glennan (1996) notes, discovering mechanisms is a gold standard for establishing causal connections. A number of philosophers developed new understandings of varying types of causal relationships in which mechanisms have played an integral role. From Hume, to Kant, to logical empiricism, regularity theories, counterfactual theories, causal process theories, probabilistic theories, to agency and interventionist theories, scientists and philosophers have attempted to identify the background and nature of how causal forces operate in the natural and social worlds. A general point to consider here is that discursively, the term mechanism has grown to be widely used as a descriptive word that would account for some form of causal operation, event, or tool. The statement I posed in the introductory chapter of this book, “what’s the mechanism” symbolizes the descriptive nature of the term and how it is implemented in the English language. On the other hand, within the widening and increase in frequency of the usage of the term mechanism, scientists and methodologists have debated on what the correct or most plausible way to implement mechanistic explanations should be.

Recall that the Humean view of fundamental causation stated that two events will lead to some cause. X must bring about Y and importantly, Humean causality may be law-based or stochastic as long as causes precede their effects (Chatterjee 2013). Hume assumed that the world is comprised of loose and separate distinct entities and existences that have no necessary connection with one another. For Hume, causal inference entailed that philosophers and scientists must discover causal relations rather than assume inference is a simple consequence of causality. Causal relations thus depend on the experiences of constant conjunctions between conditions (Garrett 2014). Given these assumptions, Hume's contributions and his greater critique of causality led to major shifts in how causality was understood with reference to our ability to make inferences about causal relations (Beebee 2006). The critique launched by Hume led to an expansion of empiricist inquiry (and aforementioned challenges by Kant, see chapter two). Of those who took on a Humean view of causation, empiricists of this persuasion (such as John Stuart Mill or John Mackie) sought to "demystify causation" (Stathis 2002). In today's debates, most who consider themselves as Humeans adopted regularity views of laws (RVL) which assume that laws of nature constitute empirical regularities (Stathis 2002, 8).

On the other hand, a framework put forward by Carl Gustav Hempel led to a trend of scientific inquiry that later would be referred to as the covering-law model of explanation. In the covering law model, there are three domains: 1) the deductive-nomological (D-N) model based on deductive arguments and premises that include universal (deterministic laws along with statements of particular conditions; 2) the inductive statistical (I-S) model which is based on arguments whose premises include statistical laws; 3) the deductive statistical (D-S) model which is based on deductive arguments in which statistical laws are comprised of comprehensive statistical laws (Sarkar and Pfeifer 2006, 276). These variants of the covering-law based explanation are specifically premised on the notion that the explanandum is the conclusion of an argument whose premises contain at least one natural law (Sarkar and Pfeifer 2006, 276). The different domains of the covering-law model have been adopted widely throughout the social sciences which are geared towards observational statistics rather than experimentation. Throughout various forms of scientific applications, mechanisms started being placed somewhere in between the explanans and explanandum and were assumed to operate as mediating forces. To date, there have been dozens (if not hundreds) of different interpretations of mechanisms across both natural and social science. Conceptions of what constitutes a mechanism traditionally have developed in both natural and social sciences (Craver et al. 2000).

Mechanisms in Scientific Explanation

Stathis (2002) argues that Hume could not identify or see the conceptual link between cause and effect. Hence, other philosophers of science (such as Mackie; Salmon; Dowe) have since contributed to theory on these links through positing that mechanisms connect cause and effect. Mackie (1974) contended that a causal mechanism is dependent on the causal order between conditions and that if X causes Y, then it must mean event X happens prior to event Y. Additionally, Mackie argued that mechanisms could be the missing link between individual cause and effect that was absent in the regularity theory of causality. Mackie's notion of causal mechanisms falls under basic laws of working, argues Ehring (1997), and the qualitative persistence in Mackie's framework grounded in spatio-temporal continuity can provide a basis for sorting through different causal orderings in a given inquiry (Ehring 1997, 117). Along these lines, Mackie (1965) conceptualized what he referred to as an INUS condition which is a cause that "is known to be an insufficient but necessary part of a condition which is itself unnecessary but sufficient for the result" (Mackie 1965, 245). This approach has been related to Mill's notions of sufficiency and necessity, and has seen espousal in cross-disciplinary research on sufficient and necessary conditions.

Another pioneering modern philosopher of science is Salmon (1994), whose work differentiated two forms of explanations–the etiological explanation and the constitutive explanation. The former type of explanation is based on accounting for an event or outcome through observing antecedent processes and conditions. The latter type of explanation, in contrast, features causal mechanisms and exhibits what Salmon refers to as an internal causal structure of explanandum. As subsequent sections of this chapter will reveal, these two modes of explanation are often not adhered to and tend to get blurred in social scientific inquiries that feature mechanistic approaches. This largely has to do with the fact that both explanatory modes are bedimmed and can intersect when applied to empirical phenomena (Salmon does acknowledge this), yet Salmon also argues that both forms of explanation can be applied in distinct formats. For example, in Salmon's second explanatory framework, reference is given to natural phenomena–Salmon gives the example of an archaeological inquiry into an old site in Alaska and notes that the site of inquiry is representative of an etiological explanation because one need not look at the causal constituents of a bone that is found in the site as we know that it is a bone. Yet in the social world, such static material phenomena never form the primary units of analysis because societal trends, groups, institutions, individuals, cultures, discourses, etc., are much more contingent and fluid and hence, require a different approach. Even though most contemporary proponents of mechanistic explanations in the philosophy of science do tend to agree that Salmon's etiological and constitutive aspects are often simultaneously at play, constitutive explanations are more widespread and emphasized in the "new mechanistic" philosophies (Chao, Chen and Millstein 2013, 153).

The profound historical upsurge of mechanistic usage in both discourse and scientific inquiry has yet to slow down, and there are vast differences in how scholars have conceptualized 1) what a mechanism is and 2) how it functions. Hitherto, a large number of different interpretations of mechanisms have been theorized across both natural and social science. If we are to understand the characteristics of these differences, one must consider that often, the ontological assumptions held by natural scientists differ than those held by social scientists. To make matters even more difficult, ontological assumptions even vary between different social scientific fields, for example between sociologists and political scientists. Even in light of these differences, mechanistic usage is more consistent (and arguably more credible) in the natural sciences when compared to the social. A critical component inherent to phenomena in the natural world has to do with the fact such phenomena do not have any self-conscious properties–microbes, bacteria, elements, viruses, plants, animals, etc., possess few conscious properties when compared to human beings. This not only sets the natural world apart from the social, but it makes the latter arguably more complex than the former. The unchanging and stable condition of natural reality (e.g., as measured in either physics, chemistry, or biology) separates the study of natural phenomena from the social. What's more, human beings are also apart of the noumenal realm, whereas no other entities, animals, or bacteria can transcend their phenomenal ontology.

For these reasons, most sociological and political phenomena do not transpire in as systematic or in as regular fashion as in the natural world. Hence, it is inaccurate to assume that the social world and the concepts that are constructed to analyze social reality are formulated with the same material and types of units of observation as those in the natural world. These differences pose ramifications for the scientific deployment of the mechanism. Our understanding the social world entails that we must engage with and attempt to discover our profoundly deep-rooted ability to practice cognition and exercise conscious awareness. In contrast, in the scientific study of the natural world and its phenomena, our power to engage in precise measurement and experimentation has resulted in very fine grained and detailed discoveries of outcomes, processes, and events that are materially constant and unchanging across most of space and time. This lack of contingency has had profound consequences not only for the identification of mechanistic forces and natural causality, but has enabled the natural sciences to lend themselves to producing more feasible results and predictions that are often supported by mechanistic explanations.

The following explanatory examples are drawn from mechanistic usage in natural science. Subsequently, the relative explanatory ineptitude of mechanistic approaches in the social sciences will be elaborated upon.

In biology and chemistry, causal-laws are far less frequent than multifaceted causal processes. The latter exist on very detailed planes and levels such as the molecular level. Mechanisms are applied to investigate the intricacies of processes such as how NOx (nitrogen oxide) arises. This process features the combustion of molecular nitrogen and oxygen and arises once a specific temperature is reached, resulting in three types of mechanistic interactions known as the Zeldovich or the Thermal NO mechanisms – leading to the eventual emission of NOx (Zeldovich 1966). Here mechanistic reactions are stable and will cultivate NOx production. Another mechanistic example drawn from natural phenomena can be found in the citric acid cycle. The Citric Acid Cycle, commonly referred to as the Krebs Cycle, is an eight-step series of chemical reactions that are fundamental as they are found in all aerobic organisms. Occurring in the mitochondria, acetate that is derived from everyday food sources (proteins, fats and carbohydrates) is converted into ATP (adenosine triphosphate) through mechanistic interactions that produce the end result of chemical energy. The latter makes up nearly all of the energy used by human beings.

The Krebs cycle was discovered initially by Albert Szent Gyorgyi and then by Hans Adolf Krebs (Spector 1956). It has served as an empirically factual framework that captures how food is converted into energy. Given the sheer size and complexity of the eight step Citric Acid Cycle, only one of the reactions of this cycle will be chosen to help illustrate the relative steadiness of mechanistic interactions in the natural world and in natural processes. To interpret how this this cycle transpires in organisms we must imagine a circle-like process that begins with one stage and moves to the next–clock-wise to the finish. Another way to conceptualize the citric acid cycle is to picture it as a carousel wheel which contains eight different mechanistic reactions (each reaction substitutes a seating cart or booth in the carousel wheel). Beginning with the first reaction, the element of Acetyl-CoA combines itself with an organic compound by the name of oxaloacetic acid (a metabolic intermediate) which leads to an interaction with H20 (water) and the eventual formation of citrate. Citrate is a necessary component for the next mechanistic interaction in the cycle to begin.

Figure 2. The Krebs Cycle

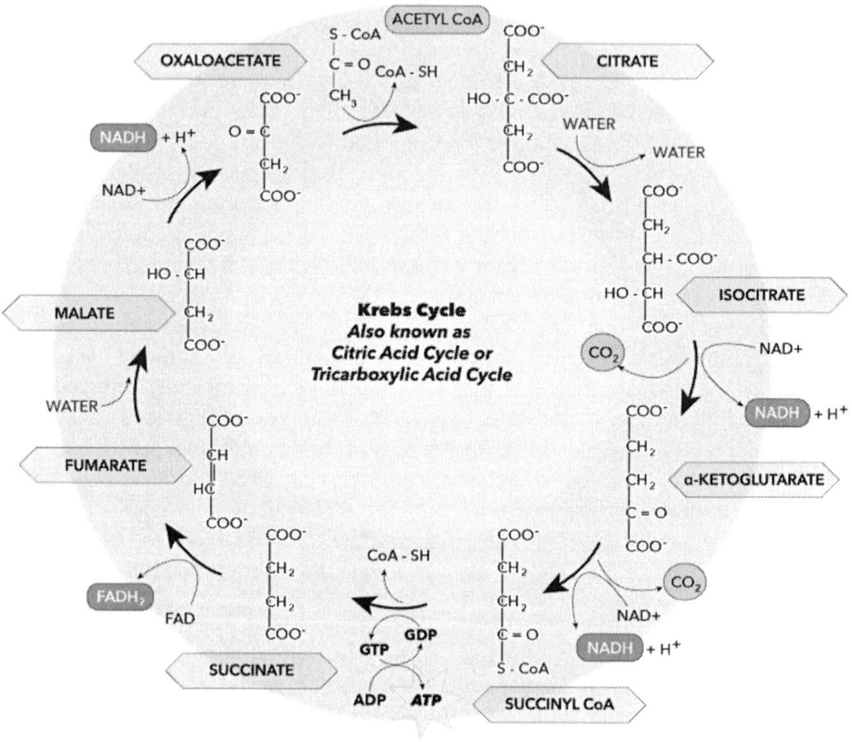

(Taken from Humphries 2022)

These mechanistic interactions exemplify a constant state of causes that take place in a common biological phenomenon. The mechanistic interactions in the Krebs cycle occur in exactly the same way, in an indefinite pattern, across time in the human body. The ontology of our social world however, is very disparate when compared to the natural world because processes that occur in the social world are exceedingly contingent. Another example from the natural sciences can elucidate the salient explanatory power that mechanistic explanations can have in enabling us to understand important outcomes can be found in the study of the human body. For instance, Van Gaal et al. (2006) explore the link between obesity and cardiovascular disease through a mechanistic approach in their study published in the respected journal, *Nature*. This study, like many in the medical sciences, relies on a mechanistic approach because it would be nearly impossible to be able to account for a phenomenon such as cardiovascular disease without understanding the intermediate forces that lay between one of the main causal conditions (obesity) that has been observed to be correlated with the outcome across both space and time. The more obese a person is, the likelier he/she will develop cardiovascular disease. Yet this does not mean the correlation between obesity and cardiovascular disease is necessarily causal in nature because on its surface, it only implies causation is going on without identification of any underlying mechanism.

This is the particular domain in which mechanistic explanations in the natural sciences shine brightest. For example, Van gaal et al. (2006) identify a number of different mechanisms that link obesity to cardiovascular disease such as insulin resistance. Individuals that are overweight, especially those that have a body mass index (BMI) of upwards of 28, are especially at risk from suffering from insulin resistance which is a condition that disables cells from adequately responding to the hormone insulin and in turn, decreases an individual's ability to efficiently store glucose (resulting in high blood sugar which damages arteries among other key components of our biology). Another mechanism is identified within the accumulation of visceral fat. High amounts of visceral fat storage will result in adverse serum lipid levels and changes in small dense LDL particles (plaque particles that tend to break off when individual experiences heart attacks and strokes). Specific adipokines that may originate from non-adipose cells in adipose tissue–macrophages–are more common in individuals with visceral fat (Van Gaal et al. 2006, 877). Further, another mechanism is elevated leptin levels that occur with the condition of increased bodily obesity. Leptin plays a role in metabolic syndrome and is said to enhance platelet aggregation, arterial thrombosis, angiogenesis, and also impairs arterial distensibility. The final mechanism linking obesity to cardiovascular disease is oxidative stress which contributes to decreased endothelial function. Obesity results in greater oxidative stress within the body due to excess energy being stored and in turn, this increases systemic inflammation and results in greater generation of LDL cholesterol.

Below, the different mechanistic processes and pathways that link obesity to cardiovascular disease are visualized as taken from Van Gaal et al.'s (2006) study.

Figure 3. Mechanisms Linking Obesity with Cardiovascular Disease

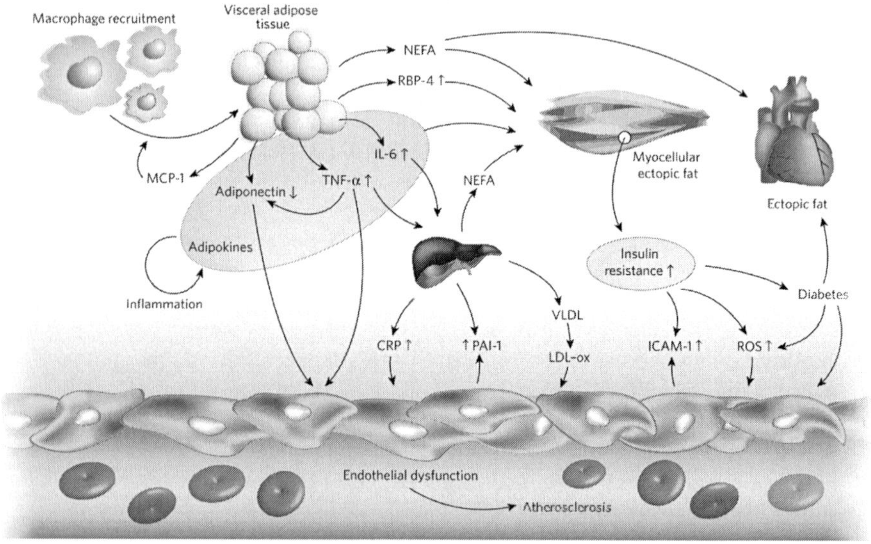

(Taken from Van Gaal et al. 2006, 878)

The authors additionally identify more mechanisms that are associated with the general mechanism of insulin resistance. They refer to these as correlates of this mechanism that might be involved in adverse left ventricle (of the heart) remodeling and dysfunction. These include: increased cardiomyocyte lipid accumulation; induction of ROS and apoptosis; generation of non-enzymatic advanced glycation end products; altered myocardial protein degradation; insulin-like growth factor-mediated effects; altered matrix remodeling; altered vascular compliance; sympathetic activation; increased renal sodium reabsorption (Van Gaal et al. 2006, 879). What's more, each mechanism can be represented by data that can be tested which brings us to the current gold standard in contemporary scientific inquiry: the double-blind randomized control trial (RCT). This type of experimental methodology is based on having two different treatment groups that that are randomly selected for treatment. One of the groups will be administered the treatment while the other will serve as the control group but will not be aware that they are the control. Often, the control group will be administered a dummy treatment (placebo). Through such an approach, specific mechanisms and mechanistic activities can be tested.

Comparatively superior forms of experimentation in the natural sciences have enabled researchers to make truly groundbreaking discoveries that have changed the fate of humanity across nearly all realms our existence. In the social sciences however, experimentation is still relatively new and is fundamentally different. Experiments are most common in the discipline of economics or political economy–where researchers make participants play economic games to assess outcomes such as altruism, selfish behavior, and cooperation. The clear divide in the effectiveness in experimentation between the natural and social sciences has resulted in the former being much more fruitful. Ultimately, the most prominent differences between mechanisms as conceptualized and identified in the natural versus social worlds boil down to and can be accounted for by the contingency of social causality. Social causation is contingent because it is dependent upon the meaning of different concepts, social practices, and organizational patterns that are present within a given historical period. These different factors and characteristics are not static in the same way an element, chemical or any biological entity is in the physical world. This does not imply that there exist no regularities or patterns in the social world, but rather, the contingency of the social world makes the application of mechanistic frameworks arduous and often leads to unimpressive results.

Mechanisms in the Social Sciences

The aforementioned Krebs cycle exists and transpires in the human body–making it a stable and verifiable empirical phenomenon that is measurable across time and space. Yet the many processes and causal relationships in the social world cannot be labeled or considered in such a manner. In the study of social reality, causation is highly contingent and according to some is even bound to social construction (Glynos and Howarth 2007). Social scientists have acknowledged that problems exist in terms of what we consider to be the contingency of social reality–as one political scientist put it, “ontology should not be defined as the world that actually exists but instead as the world as political scientists assume it to be”

(Stanley 2012, 95). Most social scientists are in agreement that mechanisms matter, but they also disagree on what mechanisms are. In the most authentic sense of the word, semantic change (and one could even say, semantic contingency) is very much present if one steps back to observe methodological debates that have arisen over the course of the last few decades in the social sciences. As stated in arguably the most widely read social science methodological textbook to date, "some scholars argue the central idea of causality is that of a set of causal mechanisms posited to exist between cause and effect. This view makes intuitive sense: any coherent account of causality needs to specify how the effects are exerted" (King, Keohane and Verba 1994, 95).

Yet there have been few (if any) analyses carried out that have resulted in the identification of mechanisms that are comparable to the earlier noted, stable process-based examples. Few social scientists would go as far as to claim they can establish causality in their analyses (correlation does not mean causation!); hence, many tend to rely on causal inference or on mechanistic explanations or a combination of both. Investigating how a given condition X influences Y is very difficult based purely on observational data and accompanying statistical analyses where the ostensible problems of external validity, endogeneity and generalizability are common. Contrariwise, small to medium-n analyses often get criticized for inherent problems of limited diversity and sample size. Because of the endemic nature of these problems, some of social science's most prominent methodologists called for cross-disciplinary turns to be made to mechanistic explanations. In the 1980s, Jon Elster issued a "plea for mechanisms" due to what he referred to as wide spread prediction failure in the social sciences. Elster's plea grew to be a very influential work (Elster 1989; 1998). Opening up the blackbox or "what's the mechanism?" are common phrases one will come across when speaking about causal relationships or processes. Such phrases seek to get us closer to understanding the underlying forces behind causal relationships in both observational and experimental settings.

Since Elster's plea however, there have been dozens of differing interpretations put forward regarding what constitutes a mechanism, and even with these different conceptualizations, no conclusive idea of a mechanism can be agreed upon. In fact, Robert Mahoney once identified over 20 different definitions of what constitutes a mechanism in the social science. Even with such discrepancies remaining unresolved, a sense of unification has arisen around mechanisms and as analytical tools, many believe mechanisms are helping to advance knowledge by turning attention to the specification of mediating steps in pathways that are at play in causal processes (Gerring 2010). Philosophers of social science have adopted mechanistic approaches and many agree with the claim that both natural and social sciences are comprised of systems and mechanisms are considered to be processes within these social systems (Hedström and Swedberg; Bunge 2004; Elder-Vass 2010). Another important point to consider here is that the rise of mechanistic use has overlapped and occurred with the increasing use of mixed-methods or what some refer to as multi-method research.

With these tendencies in mind, as this chapter has already hinted, there are substantial deficiencies in the usage of mechanisms in much of social science and specifically in political science. These have to do with theoretical mis-specifications and a general lack of conceptual reflection over how things are defined and operationalized. Mechanisms are assumed to exist and function in all social processes just as physical laws and regularities operate in the natural world. On the other hand, in some branches of sociology, mechanisms are non-law like entities that produce a given outcome regularly. Yet, with many varying conceptions, mechanisms are frequently referenced without adequate differentiation of the ontology of causation in the natural world when compared to the social. In other words, mechanistic forms of explanation are usually based on arguments that fail to take into consideration or mention social ontology. What is considered to be a "system" in the social world is not equivalent to a system in the natural world. More recently, some political scientists have called for mechanisms to be identified and related to the context in which they operate in (Faletti and Lynch 2009), while others (Grzymala-Busse 2010) have stressed the topic of temporality regarding mechanisms and causal processes.

A plausible approach was put forward by Weinert (2014) who contends that social mechanisms are based on human agency and hence must be made distinct from physical or biological mechanisms. In an intriguing analysis based on the exploration of the abolition of the 19th century slave trade Weinert argues that,

> "In the case at hand, the causal patterns were not triggered under 'unknown' conditions and the consequences were probable rather than merely indeterminate. These considerations suggest that a social mechanism cannot be employed without knowledge of the antecedent boundary conditions, under which it is supposed to occur. This result is due to the fact that social mechanisms are based on patterns, trends, and human

agency, rather than lawful regularities, as in natural mechanisms" (Weinert 2014, 20).

These critical points are worth scholars' and students' consideration because they entail that there exist a fundamental set of differences between natural and social causation, even if both are part of the phenomenal realm. For instance, in a causal process that is part of the natural world, such as the Krebs cycle, mechanistic interactions represent three out of eight requisite reactions that must arise for human beings to produce vital energy from a given food source. These interactions exemplify the stability and unvarying state of all mechanisms and processes within biological phenomena. Such reactions and interactions are also observable on a molecular level. In contrast, the ontology of our social world is unlike that of the natural because it is highly contingent and the meaning of different concepts, social practices and organizational patterns within a given state and historical period are not static in the same way an element, chemical or any biological entity is stipulated in the physical world. This is precisely why Weinert's (2014) recommendations are plausible–antecedent factors and conditions that underlie a given set of explanatory variables must be considered if we are to identify the operative nature of causation in the social world.

With that being said, this does not imply that there exist no regularities or patterns in the social world, regularities can be identified and causality does indeed exist. Events are brought about by complex, and often non-linear causes that are historically contingent. One causal process may be active in a given qualitative context at one point in historical time, while simultaneously it may not be active in other contexts in that same historical time or it may even be become active in a different context at a different historical time. There is a lack of causal uniformity in the social world and because of these complexities, over the course of the last near three decades of implementations of mechanistic explanations in the social sciences, there have been some plausible accounts produced by scholars and some not so plausible accounts. For instance, the following study may be considered to be a cross-breed between economics and political science and is among the most cited in research on what scholars refer to as the "resource curse." Humphreys' (2005) "Natural Resources, Conflict, and Conflict Resolution: Uncovering the Mechanisms" set out to "catalogue a series of mechanisms that may link natural resources to conflict onset and duration" with use of econometrics (Humphrey 2005, 5). For decades, scholars have adopted highly innovative methodological approaches (all of which are statistical) to study the cross-national puzzle of the resource curse. These lines of inquiry have sought to answer why countries rich in natural resources tend to experience adverse political, economic, and social outcomes?

The primary independent (i.e., explanatory variables) in this literature revolve around continuous measurements of resource wealth as gauged by observations on oil/petroleum, precious metals, as well as gems/stones. Additionally, as common in research on political instability, studies on the resource curse frequently include measures on the level of democracy/autocracy for each country in a given data set as well as whether the country has mountainous terrain. Humphreys argues that scholars have missed out on the central dynamics relevant to this phenomenon and that, "the problem is that unless we understand the mechanisms linking resources to conflict, the advice of conflict scholars will be of limited use to the policy community" (Humphreys 2005, 509). In his empirical analysis, Humphrey's identifies nearly a dozen mechanisms–some from his own inquiry and some from previous literature's' findings. These include the greedy rebel's mechanism, the greedy outsider's mechanism, the grievance mechanism, the feasibility mechanism, the weak states mechanism, the sparse networks mechanism, the military balances mechanism, the fragmented organizational structures mechanism, the possibility of pork mechanism, the domestic conflict premium mechanism and finally the international conflict premium mechanism. The overall goal for Humphrey is to statistically test these mechanisms and identify which mechanism holds the most explanatory power for accounting for the outcome of civil war duration (coded dichotomously).

The usage of mechanisms in this study can be thought of as an exemplar of the poor utility attached to mechanistic explanations in the social sciences. In Humphrey's analysis, the classification of a "greedy outsiders' mechanism" is meant to explain and capture situations where incentives for third parties are influential getting actors to encourage and support civil conflicts in a resource rich country. Indeed, when such components are prevalent in a given setting, it is plausible to assume the likelihood of conflict will increase. However, there is no theoretical explanation given by the author as to why the described phenomenon is a mechanism and not a set of variables or even a larger process. This demonstrates a common fault inherent to mechanistic explanations in the social sciences in that the term mechanism gets used and the concept gets applied, but the conceptual basis of such an application is totally absent of awareness of lack of causal uniformity. Furthermore, the presence of a third outside party that materially supports conflict in an unstable state is a phenomenon that empirically has occurred and is significant, yet it is not possible to categorize or operationalize this phenomenon into a mechanism as it includes more than one set of actors, structures and time periods that are simply left vaguely unidentified. The mechanism itself contains enough dynamics to formulate a possible research question, a statistical model or a political process model.

Humphreys' "weak states mechanism" is also problematic. Here states that do not rely on heavy taxation of their citizenry and rather rely on revenue from possession of their state's natural resources are categorized as a mechanism. To conceptualize as broad of topic such as a "weak state" into what is then considered a mechanism is an erroneous epistemological maneuver given this observation is an observation of a final outcome to what most likely was a long-fought political battle between relevant groups and actors. Weak states with poorly functioning bureaucracies indeed are necessary to consider when trying to answer this particular research question yet categorizing a mechanism of a weak state is a callously vague methodological decision. What is the substance of explanation in this case? That a weak state is prone to greater social conflict because of a mechanism of it being a weak state? Neither of the mechanisms identified by Humphreys can possibly be categorized as a distinct mechanism that acts as a force within a causal process but rather these mechanisms contain a stuffing of complex social phenomena into analytical variables and observations that fail to consider causal uniformity as well as relevant spatial and temporal dimensions that are relevant to consider for the phenomena under attention.

More problematically, these mechanisms have no carry-over into other historical eras and are precisely parasitic upon the time periods that have enabled them to exist. Humphreys brings up the following explanation of the domestic conflict premium mechanism, "groups that benefit during conflict may prefer to fight than to win and therefore act as spoilers to peace processes. Natural resource endowments are associated more with greed-inspired rebellions" (Humphreys 2005, 516). In other words, fighters in such situations might not have "interest" in the success of negotiations. What this mechanism is premised on is an epistemological assumption that in turn drives methodological choice. The ontology of the human being is determined by an epistemological base that itself is dictated by the homo-economus model or in other words, that of a rational actor. Instead of generalizing about "groups that benefit during conflict" it is worthwhile to explore potentially greater forces at play here. Is it that human beings (whether alone or in groups) are so determined to seek profit that they will engage in warfare relentlessly even if there is a chance of a compromise? What types of groups are these and how is it possible to generalize Humphreys' assertion onto other contexts or historical eras? Without answering these questions, the explanatory worth of aforementioned mechanisms that underlie the resource curse is dubious.

The preceding critical analysis of Humphrey's work is not meant to be demeaning to the authorship of the study, but rather, is meant to elucidate a commonality that is inherent to highly noteworthy publications that adopt mechanistic explanations in the social sciences. Lack of conceptualization surrounding the temporal, spatial, and overall substance of what constitutes a mechanism is problematic throughout many social scientific analyses. These drawbacks inspired me throughout early work during my academic career in which I utilized mechanistic approaches in combination with a discourse theoretic line of inquiry to investigate the phenomenon known as "repression backfire." In the strategic study of nonviolent civil resistance, the phenomenon of backfire is significant. During a given series of protest-state interactions, government sponsored coercion can backfire on the repressor(s). The latter processes have been referred to as either "political jiu-jitsu" or repression backfire/lash (Sharp 1973; Hess and Martin 2006; Martin 2007; Anisin 2016a; 2016b; 2018). When state repression and severe violence are exerted towards a large anti-government movement, the action(s) can spur public outrage, greater mobilization, viral media circulation, NGO shaming, international condemnations, security force defections, regime transition, identity dislocations, among other outcomes (Sharp 1973; Sutton et al. 2014).

Repression backfire is considered as a positive outcome for social movement activity because it enables protesters to keep mobilizing and to also reap benefits of new asymmetric support as well as potential sanctions getting placed on the repressor. There are a large number of historical cases involving protest-state interactions which led to repression backfire such as the 1819 English Peterloo massacre, the 1897 Lattimer Mining massacre in Pennsylvania, USA, the Russian Bloody Sunday massacre of 1905, the 1907 Chilean Santa Maria School massacre, the 1912 Russian Lena mining massacre, the 1919 Indian Amritsar massacre, the 1953 East German workers strike massacre, the 1960 South African Sharpeville massacre, the 1989 Georgian Tbilisi massacre, the 1991 East Timor massacre, among many others (Anisin 2019). Backfire has been argued to be likely to arise under certain configurations of conditions. For example, in the most prominent works in the study of nonviolent civil resistance,

Sharp (1973) argued that nonviolent direct action provides vital incentives for individuals to not only join protests, but nonviolence, when adopted as a publicly observable protest tactic, can force concessions out of the state. One of the ways that this can occur is through processes that arise when nonviolent opposition movements suffers repression.

Hess and Martin (2006) theorized two necessary conditions that they argue must be in place for state repression to be able to backfire. Even though their analysis did contain the identification of antecedent conditions which fulfills the aforementioned logic voiced by Weinert (2014), it nevertheless did not include the identification of mechanisms. The two conditions postulated by Hess and Martin (2006) are as follows: there first has to be media and network capability which can facilitate and circulate information about a given repressive act. Secondly, there must be an audience(s) to perceive the incident as unjust. A quantitative analysis by Sutton et al. (2014), found support for this hypothesis. Here it was argued that social movement communicational infrastructure heightens a nonviolent opposition movement's ability to be able to overcome state repression. The logic here entails that activist-created media institutions give individuals better chances to communicate (together and with third-party audiences) in the wake of severe repression. My engagement with this literature and the phenomenon of backfire began with the context of the first Russian Revolution (1905). In Anisin (2014), I investigated a massacre that occurred in the context of the Tsarist empire inflicted upon a protest that was organized by an Orthodox priest in the latter portion of 1904. In January of 1905, this mass protest marched to the Tsar's winter palace in St. Petersburg (then capital of the Russian empire), and was met by cavalry who murdered dozens of unarmed and nonviolent civilians. The organizational campaign that went into this protest was significant as leadership sought to gather a heterogeneous collection of social demands from different religious and socio-economic groups in society. I argued that the protest represented and equivalized force – a force that was symbolic of the "people" that the Tsar represented. Evidence to support these claims was provided through analysis of the petitions that protest leaders put forward to the Tsar along with slogans that marchers were carrying on the day of the massacre.

The figure below illustrates the mechanistic process model that was put forward to account for repression backfire in the Russian Bloody Sunday massacre of 1905.

Figure 4. Process Model of Repression Backfire 1

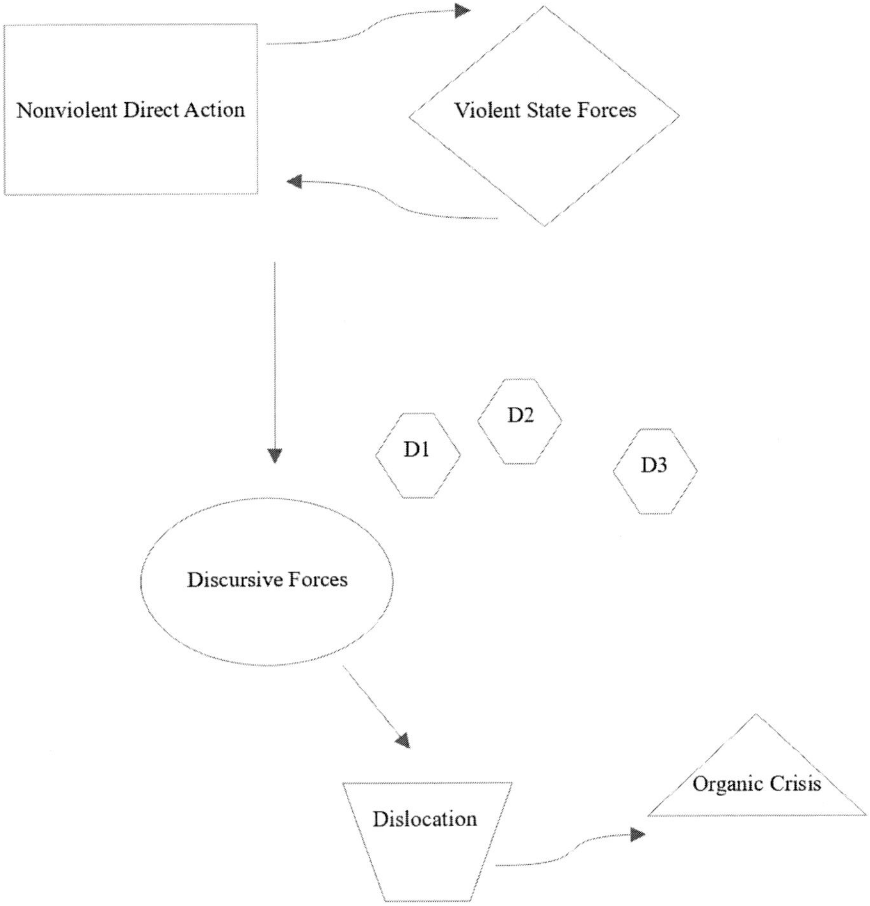

(Taken from Anisin 2014)

The first components of the process model involve the interaction of nonviolent direct action and state repression. From this interaction the process model moves into the component labeled discursive forces which alongside it has three discursively rooted mechanisms. As described,

> "The purpose of a discursive force is to precisely explain how transformative events and general social change may arise as a result of repression inflicted upon nonviolent civilians via the concept of dislocation. Here the

goal is to analyze specifically (1) how the contingency and antagonistic relationships behind the political discourse of a given regime may get revealed to populations and (2) the ways in which the fantasmatic grip a regime has on its subject may weaken due to dislocation. This requires accounting for and mapping out the logics that underpin the regime under examination" (Anisin 2014, 656).

The three discursive mechanisms include the following: 1) viral media response; 2) elite updating – in which elites and political leaders change their preferences about their loyalty to an incumbent regime; 3) public outrage – which captures how audiences get outraged and experience different emotional impacts after hearing about or witnessing civilians getting murdered by government forces (Anisin 2014, 656). The process model explains how repression backfired and ruptured Tsarist hegemony, and the analysis of the study is situated in a deeply rooted discursive assessment of the qualitative dynamics inherent to the discourse of Tsarism and the political and social logics of the Russian empire of the late 19th century.

In a subsequent study, I engaged in further theorization on the conditions and mechanisms of repression backfire through exploring a case of nonviolent protest that arose in the Turkish 2013 Gezi protests which at the time, was the largest protest to have occurred under the Erdogan government. In this context, a small case of collective action in an urban park arose to challenge a governmental-led policy of demolition of a historic park in the center of Istanbul. The question I attempted to answer was under what conditions can a small instance of protest that is met by police repression diffuse into a larger general protest movement that numbers in the hundreds of thousands or millions. The analysis identified a number of antecedent conditions that built on previous research (Hess and Martin 2006; Anisin 2014), including non-hierarchical protest; urban/spatiality; priming; event ambiguity; nonviolence; and media institutions (Anisin 2016a). I argued that when state repression gets inflicted in a context of protest that features these conditions, repression backfire is likely to follow.

Figure 5. Causal Process of Repression Backfire

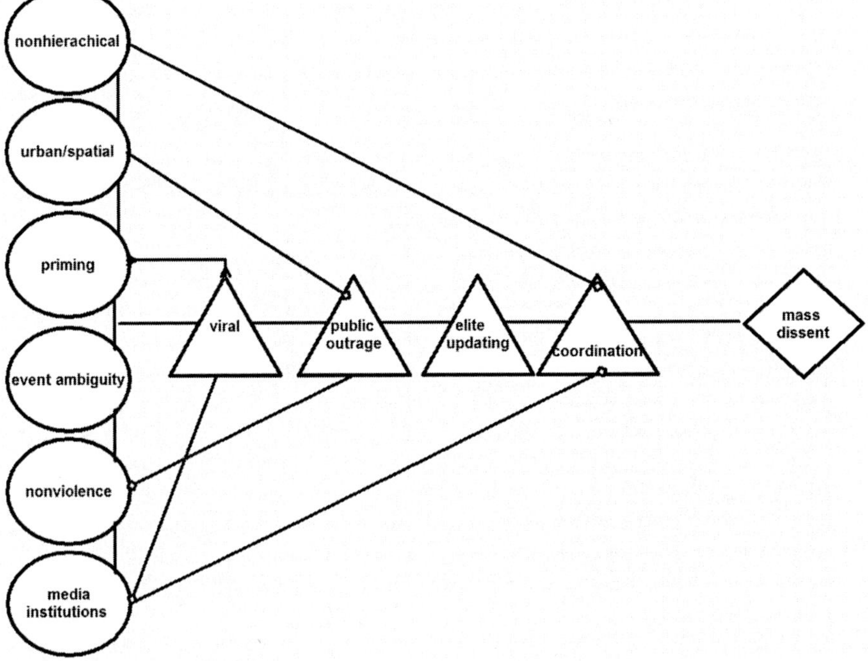

(Taken from Anisin 2016a)

The six antecedent conditions then lead to the aforementioned three mechanisms identified in Anisin (2014), viral effect, public outrage, and elite updating. In addition, a fourth mechanism is identified called coordination which leads to protesters (as well as observers) spreading information about the act of repression and collective action through variant communicational infrastructures. In cohesion, the six antecedent conditions followed by four mechanisms account for how mass dissent forms after a small protest experiences state repression and the incident results in repression backfire. This study drew on data from protest organizations and activist accounts on Twitter to compliment the analysis, which lent specific support and provided evidence for each of the mechanisms in the model and their interactions. Indeed, the two aforementioned examples drawn from my own research are not the only conceptually sound manifestations of mechanistic explanations in the social sciences, but they do nevertheless belong to a minority of studies that implement such modes of inquiry with sensitivity and cognizance about not only contextual characteristics, but also the general lack of causal uniformity in the social world along with temporal and spatial dimensions relevant to the phenomenon of interest. The latter are especially necessary to consider in future social inquiries that feature mechanistic explanations.

In summary, the grandness and progressive flare of the modern period of history was marked by a heterogeneous collection of events and processes that changed the course of human beings' relation to our environment, to ourselves, and to the cosmos. The advent of modern science, experimentation, and empiricism led to a total shift in how causality was interpreted in the phenomenal realm, in both natural and social realities. No longer were events and changes in daily life perceived to be brought about by theistic forces and powers. This chapter's assessment of these transformative changes has been supported by an account of how scientific mechanistic explanations arose in the natural sciences then were subsequently adopted and developed in the social sciences. Ultimately, the utility of mechanistic explanations turned out to be much greater and more fruitful in the natural sciences because of the technical abilities associated with experimentation, observation, and likely due to the fundamental stability of components found in the natural world. The precision with which molecules, chemicals, bacteria, and even atoms (matter) were observed by natural scientists far surpassed the observational data which social scientists started heavily relying on in the middle of the 20th century. Social scientific implementations of mechanistic explanations have by and large fallen short of what once were high expectations (of proponents of such an approach). This is not due to the faults of researchers, but rather, the lack of innovative mechanistic explanations in the social sciences is due to differences in ontological bases of the natural and social worlds. Such differences can never be overcome.

As has been demonstrated throughout this chapter, the social world is very contingent compared to the natural and empirically, social causality is qualitatively rooted with its relation to historical era(s), culture(s), and context(s). For this reason, arguably the two most transformative developments in the history of humanity arose out of the natural sciences in the just the last century of our history: the fission of the atom (nuclear power and weaponry), and artificial intelligence.

References

Anisin, Alexei. 2014. "The Russian Bloody Sunday Massacre of 1905: a discursive account of nonviolent transformation." *Politics, Groups, and Identities* 2 (4): 643-660.

Anisin, Alexei. 2016a. "Repression, Spontaneity, and Collective Action: the 2013 Turkish Gezi protests." *Journal of Civil Society* 12 (4): 411-429

Anisin, Alexei. 2016b. "Violence Begets Violence: Why States Should Not Lethally Repress Popular Protest." *The International Journal of Human Rights* 20 (7)): 893-913.

Anisin, Alexei. 2018. "Social Causation and Protest Mobilization: Why Temporality and Interaction Matter." *Territory, Politics, Governance* 6 (3): 279-301.

Anisin, Alexei. 2019. "Comparing Protest Massacres." *Journal of Historical Sociology* 32 (2): 258-274.

BeeBee, Helen. 2006. *Hume on Causation*. London: Routledge.

Bell, James. 2013. *The Spiritual World of the Hobbit*. Bloomington: Bethany House Publishers.

Bunge, Mario. 2004. "How Does it Work? The Search for Explanatory Mechanisms." *Philosophy of the Social Sciences* 34 (2): 182-210.

Chao, Hsiang-Ke, Szu-Ting Chen, and Roberta L. Millstein. 2013. *Mechanism and Causality in Biology and Economics*. New York: Springer.

Curd, Martin and Cover, J. A. 1998. *Philosophy of Science: The Central Issues*. New York: W. W. Norton & Company.

Ehring, Douglas. 1997. *Causation and Persistence: A Theory of Causation*. Oxford: Oxford University Press.

Elder-Vass, Dave. 2010. *The Causal Power of Social Structures: Emergence, Structure and Agency*. Cambridge: Cambridge University Press.

Elster, Jon. 1989. *Nuts and Bolts for the Social Sciences*. Cambridge: Cambridge University Press.

Falleti, Tulia G., and Julia F. Lynch. 2009. "Context and Causal Mechanisms in Political Analysis." *Comparative Political Studies* 42 (9): 1143-1166.

Garrett, Don. 2014. *Hume*. London: Routledge.

Gerring, John. 2010. "Causal mechanisms: Yes, but…." *Comparative Political Studies* 43 (11): 1499-1526.

Glennan, Stuart S. 1996. "Mechanisms and the nature of causation." *Erkenntnis* 44 (1): 49-71.

Glynos, Jason, and David Howarth. 2007. *The Logics of Critical Explanation in Social and Political Theory*. London: Routledge.

Grzymala-Busse, Anna. 2011. "Time Will Tell? Temporality and the Analysis of Causal Mechanisms and Processes." *Comparative Political Studies* 44 (9): 1267-1297.

Hess, David, and Brian Martin. 2006. "Repression, Backfire, and the Theory of Transformative Events." *Mobilization: An International Quarterly* 11 (2): 249–267.

Hollingsworth, Tamara, Leigh. 2013. *Martin Luther: A Reforming Spirit*. Huntington Beach: Teacher Created Materials.

Humphreys, Macartan. 2005. "Natural Resources, Conflicts, and Conflict Resolution: Uncovering the Mechanisms." *Journal of Conflict Resolution* 49 (4): 508-537.

Humphries, Katherine. 2022. Krebs cycle. Contributed by Katherine Humphries. StatPearls Publishing LLC. Retrieved from: https://www.ncbi.nlm.nih.gov/books/NBK556032/figure/article-23952.image.f1/

King, Gary, Keohane, Robert, and Sidney Verba. 1994. *Designing Social Inquiry: Scientific Inference in Qualitative Research*. Princeton: Princeton University Press.

Mackie, John. 1965. "Causes and Conditions." *American Philosophical Quarterly* 2 (4): 245-264.

Merriam-Webster. 2021a. "Natural science." In Merriam-Webster.com dictionary. Retrieved December 23, 2021, from https://www.merriam-webster.com/dictionary/natural%20science

Merriam-Webster. 2021b. "Social science." In Merriam-Webster.com dictionary. Retrieved December 23, 2021, from https://www.merriam-webster.com/dictionary/social%20science

Norkus, Zenonas. 2005. "Mechanisms as Miracle Makers? The Rise and Inconsistencies of the "Mechanistic Approach" in Social Science and History." *History and Theory* 44 (3): 348-372.

Psillos, Stathis. 2002. *Causation and Explanation*. Stocksfield: MPG Books Group.

Repcheck, Jack. 2007. *Copernicus' Secret: How the Scientific Revolution Began*. New York: Simon and Schuster.

Ryrie, Alec. 2018. *Protestants: The Faith That Made the Modern World*. London: Penguin.

Salmon, Wesley. 1994. "Causality Without Counterfactuals." *Philosophy of Science* 61 (2): 297-312.

Sarkar, Sahotra, and Jessica Pfeifer. 2006. *The Philosophy of Science*. Hove: Psychology Press.

Sharp, Gene. 1973. *The Politics of Nonviolent Action: Part Two, the Methods of Nonviolent Action*. Boston: Porter Sargent.

Spector, William. 1956. *Handbook of Biological Data. National Research Council*. Committee on the Handbook of Biological Data.

Sutton, Jonathan, Charles R. Butcher, and Isak Svensson. 2014. "Explaining Political Jiu-Jitsu Institution-Building and the Outcomes of Regime Violence against Unarmed Protests." *Journal of Peace Research* 51 (5): 559–73.

Van Gaal, Luc F., Ilse L. Mertens, and Christophe E. De Block. "Mechanisms linking obesity with cardiovascular disease." *Nature* 444 (7121): 875-880.

Watson, Peter. 2005. *Ideas: A history of Thought and Invention, from Fire to Freud*. London: Harper Perennial.

Weinert, Friedel. 2014. "Social Mechanisms and Social Causation." *The Journal of Philosophical Economics* 8 (1): 1-24.

Zeldovich, Ya. B. 1966. *Physics of Shock Waves and High-temperature Hydrodynamic Phenomena*. Dover: Dover Publications.

CHAPTER SIX

MECHANISMS AND ARTIFICIAL INTELLIGENCE

Neon lights glow into a gray night's sky as cars line up one behind another in one of Hollywood's most iconic streets. Cafes, restaurants, and food trucks are filled with people yet something seems out of place. Seated, standing, or walking, people are out of touch, as if they are present but not fully present. Not a minute goes by until someone looks at their handheld technological device. Some talk into their devices, record themselves, while others take photos of their food. Some cars have people in them, while others are self-driving autonomous vehicles with their driver sitting in the backseat looking at a large high-definition tablet screen. Lined up behind one another in traffic, this collection of interactions supersedes its local context and is a microcosm of the state of humanity in the digital age–an age marked by a total disconnection of human beings from the earth and the natural environment and an age of deep entrenchment into technology and autonomous forms of social organization. This chapter will overview and analyze the onset and duration of our current era of digitization and technology. Just as social causality shifted from the Middle Ages into the era of modernity and the rise of science, within our current era, social causality is starting to transmute into something entirely new when compared to its antecedent configurations.

This chapter draws on Berdyaev's aforementioned framework and categorizations of historical periods as well as on Heidegger's writings on the essence of technology. Both works will enable readers to conceptualize the differences that modern technology spurred for human civilization in comparison to previous inventions and their associated changes that have been identified throughout this book. Afterwards, literature from both scientific and philosophical viewpoints will be engaged with to explore the onset and subsequent persistence of artificial intelligence. This will be followed by examples of how artificially intelligent systems and processes are currently dictating social outcomes and causality in daily life for billions of people.

The transition from the Middle Ages into the modern era of history saw the rise of natural science, experimentation, and a dominant view of social causality that was empirically deterministic in its characteristics. The latter was assumed to consist of the newfound idea that human beings could identify mechanistic activity in the empirical world through usage of their empirical senses, and that events and outcomes were no longer determined or brought about by any theistic power or force. This fundamental alteration in understanding changed the course of human history and only became more solidified as time went on. The twentieth century then saw the rise of widespread usage of electricity, radio, telegraphs, television, and then computers and the internet. The worldwide web (as classical users tend to refer to it), enabled scientists to develop the concept of artificial intelligence which then shifted from being just a conceptual set of computer applications to a full-blown enterprise that by the 2000s, became present in daily life.

While modernity led to the easing of religious determination over social, economic, political, and cultural aspects of life, by the mid-20th century, an entirely new age started to appear. This has been described as an era of post-modernity (Lyotard 1984). Post modernity also has coincided with the rise of post modernism–a vast discursive and cultural collection of parts that intersect between gender equality, racial equality, sexual equality, animal rights activism, environmentalism, among other forms of activism and thought. In what is arguably the most radical segment of this broad array of ideas and motives, transhumanism has emerged as a movement that seeks liberation from the human condition in itself–transhumanism seeks to pave way for the first post-human social order in which human beings will not be "constrained" by their biological limitations. The age of digitalized information has arisen alongside the period of post-modernity and in recent decades, information has become increasingly commoditized. At the same time, public trust of governments in much of the Western world has reached historical lows. Many have argued that we are now in an era of "post-truth"-a new period of time in human history that is marked by circumstances "in which objective facts are less influential in shaping public opinion than appeals to emotion and personal belief" (Oxford Dictionary). Individual feelings now have more weight than evidence does, and alternative facts often replace actual facts in public discourse (McIntyre 2018). Present-day issues related to this include previously unforeseen levels of distrust in government, institutions, and even distrust in science, as well as in scientifically valid (and safe) health procedures. Similar to how modernists presented arguments about the mythical character of religion and God, some post-modernists have stated that science is also comprised of myths, and that there are competing scientific paradigms that are similar to sects, contending against one another for grant allocations and public prestige.

The post-modern era is the first that contains the creation of intelligent artificial systems. Rapidly advancing technological innovation has resulted in the arrival of artificial superintelligence–a new breed of intelligence that exceeds previous cognitive capabilities. There are significant parallels between the development of the world wide web (internet) and artificial intelligence (Halpin 2004). The most common interactions that humans have with AI in today's world is in their daily usage of the internet and specifically in social networking sites–platforms that are goaded by machine learning. One need only to take public transport in any global metropolis whether it be New York City, San Francisco, London, Shanghai, Prague, Moscow, Buenos Aires or Mexico City to observe that out of every given 10 people, a super majority of them are on their smart phones, scrolling through sites that rely on artificial intelligence to curtail informational content to their users. These sites monitor their users' actions, pick up, track and record their location, previous search histories, and then from an unfathomable amount of data, they dictate what piece of information a user will be exposed to next. Users contribute to a globally connected web that in its totality represents what tech-mogul Elon Musk has described to be as a globally pervasive, conscious and alive interconnected network of machines and humans.

Accounting for Modern Technology

There have been a variety of different approaches that people have thought of in their attempt to account for the emergence of modern technology and one would not be wrong if he/she were to posit that there are different stratospheres to what we believe constitutes modern technology. Radio communication, television, digital cables, the internet, and a global system of satellite technologies (global positioning systems), have fundamentally transformed the way that human beings receive information, communicate with one another, and ultimately, the combination of modern technologies led to the formation of an artificially intelligent-dominated virtual reality. Berdyaev foresaw not only that such a collection of developments would arise in his categorization of the fourth period, but he even foresaw that there would be autonomous spheres of technological operation. The fourth period emerged in the twentieth century and is "the final metamorphosis of the realm of Caesar" (Berdyaev 1952, 48). Berdyaev begins his description of this period by placing emphasis on what he coined as the "disruption of cosmic order," which occurred when the atom was split and nuclear weaponry was developed. While many international relations experts have since placed great emphasis on both coercive capabilities and stabilizing factors associated with nuclear weapons, Berdyaev saw the creation of the nuclear bomb as a tragedy that paved way for an entirely new relationship that human beings would have in relation to nature and the cosmos.

The splitting of the atom solidified the full and total dissolution of organic forms of human organization. Berdyaev emphasized that for the first time in history, a grave new potential to wipe out lifeforms on earth arose. Nuclear weaponry and power could disrupt the universal configuration of space and the makeup of the cosmos. Because of this potential, Berdyaev saw that the realm of Caesar became "terribly augmented" with power that human beings had developed over nature. Through the usage of machinery and technology, human beings not only learned to control nature, but they continued to improve their systematic manipulation of nature for profit-based motivations. In this sense, the rise of the fourth period did not simply lead to a departure from the third, but rather, it built upon previously established marketization that emerged alongside industrialization. Once the Cold War ended and the Soviet Union collapsed, nearly all countries in the world adopted a market-based economic system in which profit seeking serves as the principal basis of economic exchange and production. This is truly a profound shift if we consider that currency was only invented several thousand years ago. As we are still in the age of economic globalization, the aim of profit-seeking does not appear to be slowing down, especially if we consider digital marketing and the abilities associated with algorithmic prediction(s) of consumer behavior.

Furthermore, Berdyaev also foresaw the rise of autonomous machinery and spheres of organization that were neither part of nature nor exclusive to human consciousness. He noted that, "machines are made from material elements taken from the old nature, but into them goes something quite new, no longer of nature, and not a part of the old cosmic order" (Berdyaev 1952, 51). In the fourth period, machines and their underlying technology would grow to have a "cosmogonic significance" (Berdyaev 1952, 51). An interesting claim was postulated by Berdyaev in that, "this is a new day of creation–or a new night. Probably night, for the light of the sun may be darkened" (Berdyaev 1952, 51). The important factor to consider here is that Berdyaev specifically classified the fourth period as a time that would experience the creation of a new type of human organization–one that was, "distinct from the organic, technics and mechanization" originally found in the industrial era (Berdyaev 1952, 47). The fourth period does not require sanctification which the realm of Caesar demanded in previous eras as this period contains the final phase of secularization. Along these lines, Berdyaev foresaw that the role of technology and machine-based production would lead to great personal difficulties for human beings because we are insufficiently adapted to handle such augmentations. The great difficulty would come with the "terrible shortening of time, a speed with which man cannot keep pace" (Berdyaev 1952, 51). He foresaw that under such a technical, machine-dominated order of life no single moment would be of value itself as each moment would become, "only a means for the next" (Berdyaev 1952, 51).

Berdyaev's ideas on technology were undoubtedly influenced by Heidegger as he engaged with his highly influential philosophical contributions that began in the late 1920s and stemmed into the 1950s (the decade in which Berdyaev passed away). Heidegger's most important work on modern technology literally probed the ontological condition of technology by posing the question of what is the essence of technology? In his essay titled, *The Question Concerning Technology* as well as in the Bremen and Freiburg Lectures (1949), Heidegger offered an account of how modern technology differs from technology of the past. The etymology of the word technology can be traced back to the Greek "technikon"–a term that also was used synonymous with techne. Heidegger associates techne with creations that arise from engaging in handicraft in making items and objects while also being illustrative of artistic practices (Anisin 2022a). From this perspective, Heidegger then argues that there is a poetic nature to techne because when it is practiced, it results in *poiesis* (Greek – "to bring forth" or bringing forth). This age-old artistic form of technology was emblematic of human beings' unique ability to create objects and items. Yet with the rise of modern industrial society, a totally new and different essence of technology emerged. It is within this new form of modern technology that Heidegger argued humans are entrapped into an ontological condition that subjugates and reduces us into specific types of almost non-human entities.

Dreyfus (1995) described this key aspect of Heidegger's argument as constituting an unintended obsession with treating our world and ourselves as resources that can be ordered in a scientific way (Dreyfus 1995, 42). This is why in the actual essay on technology that Heidegger is most well-known for, he introduced his purpose of inquiry as follows,

> "We shall be questioning concerning technology, and in so doing we should like to prepare a free relationship to it. The relationship will be free

> if it opens our human existence to the essence of technology. When we can respond to this essence, we shall be able to experience the technological within its own bounds" (Heidegger 1977, 287).

In the beginning of his analysis, Heidegger took issues with Aristotle's framework "the four causes" by asking why are there only four causes? Likewise, Heidegger scrutinized what the meaning of "cause" really can be (Heidegger, 1977, p. 290). This is where the aforementioned concept of *poiesis* comes in to play as it is articulated to be outside of the four causes and as a process that involves "bringing forward into appearance." This process thus represents the act of creating which is separate or outside of a person and the creator of the object. Since *poiesis* means bringing-forth, then the four causes are at play within this process. Heidegger then makes the proposition that not only does *poiesis* contain all four of the Aristotelian causes, but it actually takes on both cause and effect. Hence, technology is representative of something which is much grander than the commonly assumed function not simply a means to an end but it has a principal function as a way of revealing (Heidegger 1977, 294). Modern technology, through revealing and bringing-forth the value attached to techne, is conceived of through scientific thought via a process of ordering. Heidegger refers to this as Gestell (positionality) which is a concept premised on the assumption that human beings reduce phenomena to resources that are waiting to be fit into some form of technological system.

Gestell gathers different types of entities in a certain way and predispositions them into technological structures or systems (James et al. 2018). This obscures human creativity and limits our possibilities to engage in other types of being and experience which brings us to the crucial historical fact that Heidegger believed set modern technology apart from forms of the past. He viewed technology as a reflection of human beings' modern disposition to control and serve as masters of nature. For instance, in his description of manufacturing, Heidegger noted that, "The manufacture and utilization of equipment, tools, and machines, the manufactured and used things themselves, and the needs and ends that they serve, all belong to what technology is. The whole complex of these contrivances is technology. Technology itself is a contrivance—in Latin, an instrumentum" (Heiddegger 1977, 288). There is no revealing or bringing-forth in modern technology in the same way as in technology of the past. Accordingly, as noted by Heidegger, "so long as we represent technology as an instrument, we remain transfixed in the will to master it."

The analogy given by Heidegger that enables him to operationalize the old versus new can be observed in his discussion of a windmill situated on a river, “Its sails do indeed turn in the wind; they are left entirely to the wind's blowing. But the windmill does not unlock energy from the air currents in order to store it” (Heidegger 1977, 296). New forms of technology (e.g. a coal plant) arose with industrialized society and were integral to the historical emergence of such a society,

> “The work of the peasant does not challenge the soil of the field. In sowing grain, it places seed in the keeping of the forces of growth and watches over its increase. But meanwhile even the cultivation of the field has come under the grip of another kind of setting-in-order, which sets upon nature. It sets upon it in the sense of challenging it. Agriculture is now the mechanized food industry. Air is now set upon to yield nitrogen, the earth to yield ore, ore to yield uranium, for example; uranium is set up to yield atomic energy, which can be unleashed either for destructive or for peaceful purposes” (Heidegger 1977, 296).

These new conceptions of technology differ from the old “handicraft” technology, argues Heidegger. When viewed through a larger historical trajectory of human development, Heidegger claimed that the essence of technology stems to the advent of modern physics (which arose in the 17th century) and its embedded processes of ordering. The conceptual ordering that underpinned the formation of modern physics enabled science to be used for the development and efficient implementation of modern technology to master the world’s resources. With a newfound ability to dominate and control nature through extracting resources, human beings in the modern era of history are now able to unlock nature’s sources of energy, transform it, store it, distribute it, and ultimately, fundamentally alter it. These processes of extraction, accumulation, and production all take place without any particular personal attachment of human beings to the resources that are being used during production. For this principal reason, modern technology was argued to be an adverse historical development which is why Heidegger believed reliance on technology subtracted away from human potential and creativity. Accordingly, the modern technological process no longer brings forward objects, but engages in production for simply the purpose of ordering. Basing our approach to life through technology results in us treating everything as manipulative resources and carrying out our day-to-day activities in a way that prevents us from getting deeply involved in the things that we are doing (Dreyfus 1995). As subsequent sections of this chapter will reveal, lack of involvement, unfortunately, is the least worrying of problems associated with the digital age and new configurations behind social causality.

The Emergence of Computing

Nillson (2010) notably stated that AI may lack an agreed upon definition, but people that write about it and its history must have some kind of definition in mind. Nillson's own description of AI is that it represents activity devoted to making machines intelligent. Because of the broadness associated with intelligence, the history of the subject features a plethora of different fields and topics such as control engineering, electrical engineering, statistics, linguistics, formal logic, biology, and computer science–clues about what can make machines intelligent are scattered throughout all of these different disciplines (Nillson 2010). Both Aristotle and Leibniz's contributions to logical reasoning set the foundations that were later relied on when creating artificially intelligent machines through syllogistic rationale, argues Nillson. George Boole's early contributions to mathematical theory also were of great importance and later became known as Boolean algebra–a reasoning system that went on to form the basis of modern computing. Nahin (2017) puts forward a plausible argument in his book by stating that Boole was among the principal figures in history that created what later turned into the age of information. The importance of Boole's contributions is in the fact that he put forward the first logical reasoning that could be manipulated by equations which represented logical propositions, and in turn, this later turned out to be "a very important clue about the mechanization of reasoning" (Nillson 2010, 32).

As noted in the second chapter of this book, up until the 1960s, statistics (and statistical theory) was quite infantile when compared to its contemporary makeup because it was based on what many refer to as a "frequentist" logic which is in its core, static and unable to account for probabilistic changes. Thomas Bayes' contributions changed the status and explanatory power of the discipline of statistics with his theorem that later became known as Bayes' rule. This eventually led to the formation and widespread dominance of Bayesian statistics. Nillson (2010) attributes the importance of Bayes' theorem to its ability to offer us signal detection which enables us to make intelligent estimates about randomness in nature. Through being able to mathematically model "nature," Bayes' theorem enabled scientists (and mathematicians) to ascertain the probability that an event will occur based on prior knowledge of that event. If one does not have any prior knowledge of the given phenomenon or outcome of interest, he/she would logically assign a 50% probability to both the outcome of event occurrence and non-occurrence. When both probabilities are equal, then researchers will tend to rely on the maximum-likelihood method of estimation. Eventually, Bayes' rule enabled scientists to imagine how perceptions of speech sounds and visual images could be mechanized (Nillson 2010, 53).

Indeed, Booles' and Bayes' contributions to mathematical theory and statistics are not the only or sole factors that went into the underpinning of AI, many other great contributions were made by scientists such as Pearson, Fisher, Wald, Neyman, Van Meter, Middleton, Nash, among numerous others (Nillson 2010, 53). The 1950s and 1960s saw the first implementations of pattern-recognition which resembled neural network techniques. This led to a computational based ability to distinguish between different features. Pattern-recognition methods were carried out by aerospace and avionic companies that had long term budgetary flexibility due to contracts with the US Department of Defense (Nillson 2010, 105). Many of the general-purpose technologies invented in the twentieth century stemmed to governmental funding including atomic energy; commercial aircraft; electric power; computers and semiconductors; the internet; space industries–all of which represent what Ruttan (2006) describes as revolutionary departures from existing technological trajectories (Ruttan 2006). Along these lines, the military's role in developing computers, semiconductor technology, and software technology was principal as with the support of the military, fully electronic computing machines were developed after WWII.

Subsequently, markets that relied on such technologies flourished and the same time were nourished by the defense, energy, and space industries (Ruttan 2006). International Business Machine's first automatic calculator was created in collaboration with Harvard University and is but one example of a project that received governmental funding along with IBM's Naval Ordinance Research Computer or Ratheon's Hurrican (RAY-DAC) among dozens of other projects that were implemented from the post WWII period until the 1960s. As time went on, computing power increased at a rate so rapid that few could comprehend or foresee the dramatic impact that this would have on human civilization. In 1970, the number of transistors per chip in Intel microprocessors was only 4,004; each four years after the number doubled and continued to increase. In 2002, there were 256,000,000 of such transistors per chip–an amount that goes into low-end smartphones that were on sale in the 2021. The key factor to consider with relation to the development of computing and the subsequent emergence of AI is that the actual progress and evolution of these processes is bound to increase in an incremental way. Gordon Moore's notable article in an electronics magazine titled, "Cramming More components onto integrated circuits" (1965) identified a revolutionary trend which later became known as Moore's law. Moore discovered that the complexity for minimum component costs had increased at a rate of roughly a factor of two per year, and that the number of components per integrated circuits is bound to multiply in times ahead (specifically, the density of transistors will double every two years), and that no foreseeable factor will slow this increase down apart from the potential lack of availability of materials (i.e., supply chain problems) needed to create circuits and chips.

In 1980, there were 8,086 transistors in an Intel processor, by 2010, there were over 1 billion. With the growth of the global semiconductor industry, the early hegemons of IBM and Apple Macintosh remained relevant up until the time of writing this book, but importantly, throughout the last forty years, other important developments arose such as the rise of the ".com" or "www." era of the internet, multimedia computers, digital cellular phones, DVD players, electronic commerce, 2,3,4, and 5g cellular phones, cellular phones with cameras, smart phones, smart watches, wireless technology, charging, and media tablets. The application of these technologies has thus far been wide reaching across nearly all spheres of human existence including: private household services, utilities, hotel and restaurant industries, social services, education, agriculture, health services, public administration, policing, defense, financial services, construction, retail, real estate, manufacturing, among numerous other components of social organization. Perhaps the most important development of all of these aforementioned factors arose when the first internet network was established in the 1960s in the United States through the Advanced Researched Projects Agency (ARPA) via the US Department of Defense (DOD). The DOD was particularly interested in forging an architecture featuring a sophisticated communications system that would be able to function in light of a nuclear attack and nuclear warfare by a foreign adversary (which if enacted, would wipe out traditional communication systems). The Cuban Missile crisis also exacerbated the general motivations to create a new high-tech system of communication. By 1971, ARPANET was established and centers at Stanford, Berkeley, UCLA, Rand, MIT, Harvard, etc., were working together on information processing techniques to improve communication (Ruttan 2006).

The Emergence and Dominance of Artificial Intelligence

Already in 1959, computers could play simple board games and by 1997, they could play the most complex of board games–chess–beating Garry Kasparov, a chess grand-master. By 2008 AI was used by computers to recognize faces in pictures. Automated translation arose in 2010, real-time translation spoken words in 2016, and subsequently, the first driver-less cars went into (non-commercial) operation (Wooldridge 2021). Wilks makes a funny and at the same time, interesting claim in that, "Wittgenstein wrote that a philosophy book could be written consisting entirely of jokes. In that spirit, an AI book could perhaps be written now consisting entirely of snippets from the daily news making claims about breakthroughs and discoveries" (Wilks 2019, 1). At the time of writing this book, AI based computer applications are currently able to write scientific essays, stories, interpret works of art, and eventually, are going to be able to engage in what Wooldridge (2021) refers to as "human-level general intelligence." One may ask, how can computer applications that rely on artificial intelligence progress at such a rapid pace and specifically, how can they account for accounting for uncertainty in the empirical world? Bayes' theorem is the direct answer as it enables a system to able to rationally adjust its beliefs. By the end of the 1990s argues Wooldridge, AI reached maturity and this paved way for the onset of machine learning. The latter is a relatively straight forward process in which programs can compute desired output from a given input without being told a specific recipe for how to do it (Wooldridge 2021).

When Google acquired a small company called Deepmind in 2014, the large sum paid for this acquisition shocked many in the tech industry (and the overall world of business) because the company was not widely known and its services were considered to be technologically niche at the time. Yet upon closer examination, Deepmind was based on AI through its program AlphaGo (a game similar to Chess). This acquisition made the technological status of AI big news, argues Wooldridge, "In less than a decade, AI had been transformed from a quiet backwater of computing into the hottest and most hyped area of science" (Wooldridge 2021, ch. 5). As noted by McCarthy (1989), artificial intelligence is the science and engineering of intelligence machines. When deployed to carry out a given task or function, AI has continually superseded the capacity and ability of human intellect. The primary technology put forward by Deepmind is referred to as deep learning which Kelleher (2019) argues now constitutes among the most fundamental of technologies associated with AI because it can accomplish diverse sets of tasks through usage of simple calculations.

Deep learning is a subfield of AI based on the creation of large neural network models. The ultimate purpose of deep learning is to produce precise and accurate data-driven decisions (Kelleher 2019). Kelleher (2019) classifies deep learning as being in the umbrella of machine learning which itself is in the umbrella of AI. When it comes to large datasets, deep learning is the most accurate because of its ability to deal with high dimensional domains such as pixels in a photo (Kelleher 2019, 33). Deep learning is particularly suited to implement the principles inherent to backpropagation algorithms which adjust weights in a network so that each weight gets updated "in proportion to the sensitivity of the overall error of the network to changes in that weight" (Kelleher 2019, 223). One problematic development that has arisen from these advances in AI has to do with generating massive labeled datasets which then get utilized generative adversarial networks (GANs). The GAN can learn to synthesize fake images and opens the possibility of automatically creating massive labeled datasets which then get used to generate what are referred to as "deep fakes" (Kelleher 2019, 236). Deep fakes are often used maliciously and tend to be implemented in the grander scheme of fake news production as they enable swapping a given person's face onto someone else's body. We will return back to the topic of deep fakes in the subsequent section.

In 2013 the European Commission funded a massive program (1 billion euros) called the ten-year Human Brain Project that has since seen upwards of 500 scientists work across research centers in Europe with the ultimate aim of developing neuromorphic computing platforms that can simulate a complete human brain. Along with the Human Brain Project, IBM is currently running the TruthNorth chip which has a similar purpose (Kelleher 2019, 243). Elon Musk's Neuralink Corporation is probably the most significant attempt of forging a cyborgic computer-human environment to date. Established in 2016, Neuralink has been creating small devices that are meant to be inserted into the human brain. The purpose of this cyborgic development is marketed as treatment for cognitive deficits, brain diseases, and the eventual linkage of brains to computer networks. Only the childlike can honestly believe such claims. The chip that is planned to be inserted into a person's artery is referred to as a neural lace which Musk describes as a way to connect the brain to a computer system through a "lace-like appendage" that gets draped over the brain as a form of digital third layer above the cortex (Ferris 2017). The linkage of the neural lace to the human body has been described by Musk as being planned to function as some sort of roadway that connects to neurons in a person's body. The system is currently being developed to function through at least two orders of communication channels (electrodes) that are based on fully wireless communication through skin (Neuralink 2022).

Whether or not Neuralink will be successful in its attempt to build a fully integrated Brain Computer Interface (BCI) system, the attempt itself represents a current movement by information technology corporations and interests into a realm that will sooner or later result in human beings' synchronization with virtual realities that are dictated by AI. Interestingly enough, the antecedent conditions that one would imagine have to be in place for AI-driven mechanisms to function in the social world are already in place. Although we do not have microchips in our brains, most people in the developed world carry around smartphones and spend enormous amounts of time on their devices each day. In this sense, smartphones are already next to our brains, and our senses are fully engaged in digital content. A survey from 2021 carried out on over 2,000 respondents in the US revealed that 46% of people spent 5-6 hours a day on smartphones and 22% spent 3-4 hours a day (Ceci 2022). These numbers are only going to increase with each year that goes by. In terms of young people and especially teenagers, some estimates place the average time spent on smartphones and social media to be upwards of eight hours a day. This self-inflicted process of enslavement is incredibly worrisome. The more time one spends on such platforms, the likelier it is that he/she will develop adverse changes in psychiatric health including depression, anxiety, antisocial behavior, among other outcomes. The shocking aspect of this trend is that these adverse outcomes are bound to arise in people who did not have prior psychiatric problems (Riehm et al. 2019). What's more, commonly gathered data on smartphone usage tends to only include observations of time spent on smartphones, which misses out on the other device (computers), that great portions of adults spend their working day on.

Furthermore, social media and digital platforms are based on frameworks that are dictated by AI algorithms. AI determines the type of information that people receive and how it is portrayed. In the future, the power of computing will only increase and the fundamental basis of computing will take leaps that are so big they are immeasurable to our current state of cognition. The current development of quantum computers by major corporations such as Microsoft and Intel are leading to the creation of machines that will be more than one million times more powerful than all current existent computers on earth today (Vice News 2019). These new machines will be not be based on the common binary computational structure (0 – 1) that current computers function on, but instead will be principled upon superposition, in which computer decision making will function through quantum physical dynamics via the "qubit." The latter will enable computers to process while being both 0 and 1 at the same time. As one physicist notes, at 300 qubits, a computer will be able to decode more information than all of the atoms in the universe put together. It will also make computers multitask to an extent that was previously incomprehensible as each qubit will work simultaneously. The most prominent applications of quantum computing and AI that are currently active can be observed in military equipment and machines such as the US F-35 fighter jet which runs on more than 8 million lines of code and utilizes quantum computing produced by defense contractor Lockheed Martin. Quantum computing is used to verify large amounts of code and make sure the aircraft operates reliably (Weinberger 2014). Similar applications are planned to be used in businesses for mining complex data structures or in security applications such as image pattern recognition.

The Russian robotic mini-submarine called "Poseidon" was created in the last several years and there are plans for over 30 of these machines to be manufactured for the Russian Navy. With an ability to travel at speeds exceeding 100km per hour, this underwater drone is not only the fastest underwater nuclear capable weapon ever created, but it also contains doomsday plans that it can develop on its own in case of a nuclear strike being carried out on Russian soil. There is indeed irony in the name Poseidon–as this was the Greek God of the Seas but in today's world there is now a Poseidon that is an AI-based nuclear submarine drone. This particular machine is capable of articulating and carrying out and its own doomsday mechanisms in which it will either detonate nuclear blasts under water or send nuclear missiles into the air if it cannot receive updating signals (during an attack) from Moscow and other areas of the Russian Federation that have nuclear emergency warning systems. If adversaries of Russia successfully strike its territory to the extent in which there are no human beings left to operate nuclear launch systems and centers on mainland, Poseidon can create an artificial radioactive tsunami that reportedly can range from 300-500 meters in height and can engulf an enemy country's terrain for hundreds of kilometers (Fakhrutdinov 2018).

From Ancient Greek Theater to AI and Quantum Computation

Even with heavy criticism being aimed at AI in its early stages of development, by 2022, all criticisms of its utility can be considered to be null because it has superseded our collective cognition and cognitive power. Kurzweil (2005) argues that singularity is near. In the upcoming years humans will transcend their biology due to technological advancements. Kurzweil specifically projected this will occur in what he categorized as epoch five (the merger of human technology with human intelligence), which was preceded by epoch four (technology), epoch three (brains), epoch two (biology and DNA), and epoch one (physics and chemistry). The singularity principle of technology entails that in the future, technological growth will be irreversible and beyond any single person or group's control. Kurzweil was among the first to offer a comprehensive account of how this would occur through predicting that technological machines will supersede human beings in their cognitive capacity and power. Much of Kurzweil's initial argument was correctly premised on Moore's law and the idea that informational technologies indefinitely increase their technical capacity and bandwidth as time goes on. Kurzweil also argued that once AI systems supersede human beings' level of cognition, they will eventually be able to articulate their own category of ideas that will supersede anything human beings can come up with. Some chess AI programs (e.g., Stockfish or AlphaZero) have the capability to calculate more than 60 million different moves or positions per second. The AlphaZero program taught itself chess after being assigned only the basic rules of the game then it subsequently self-created strategies through reiterations of millions of games. Learning from its past mistakes, the program has been described as possessing a human-like drive to make progress and "never sit still" (Regan and Sadler 2019).

Kurzweil does not take a cautious or wary approach to these new technological developments but rather, similar to E. Musk and other tech proponents, he erroneously assumes the future will be brighter and human life spans will even be extended because of AI's projected ability to cure biological aging, diseases, detect cancer in its early stages. Again, similar to Musk, Kurzweil also foresees that the human beings' cognitive states will not be "limited" to our own mind but will be synchronized to a symbiotic cloud service (similar to Neuralink). Retrospectively, if notable early critiques of AI are considered, then it is easy to see that they were often tied to an argument put forward by Hubert Dreyfus (1972) who posited that AI and computers can never be equivalent to human beings because of our experiences in learning that we went through during processes of growing up in our lives. Since only beings that grow up can actually experience such interactions and thought processes, they develop cognitive intuition that is not necessarily premised on formal logic but is intuitive in its nature. This was a principal difference, argued Dreyfus, because it meant that AI and computers are bounded to a Boolean logic while the human mind is not. Another critical assessment of AI can be found in a more recent work (Bostron 2014), who sees AI and its future development as posing a grave danger to our existence. Bostrom (2014) argues that malignant failure modes of AI can lead to human extinction through AI making choices that are completely contradictory to the intent of their programmer(s). These developments will arise at the precise point in which AI-based machines will be able to make their own machines and programs which is why Bostrom has frequently been quoted for saying, "we humans are like small children playing with a bomb," and that, "we have little idea when the detonation will occur, though if we hold the device to our ear, we can hear a faint ticking sound" (Adams 2016). We will turn back to Bostrom's argument in this book's conclusion.

Furthermore, in research on AI, the term "mechanism" is frequently brought up with reference to different functions and factors that are inherent to this computational enterprise. At the time of writing this book, specific mechanisms akin to AI have been identified and developed by scientists. Guerin (2021) in a recent study titled, "Projection: A Mechanism for Human-like Reasoning in Artificial Intelligence," identifies a key mechanism that solves the problem of applying knowledge to challenging situations that are marked by variance. This is a mechanism that humans implement through our own knowledge, but its application in the realm of AI differs fundamentally from our own qualities due to our lack of ability to be able to draw upon near infinite amounts of data and datasets to drive decision making when faced with a set of choices or circumstances. The projection mechanism functions in a top-down manner and helps to solve problems across a range of AI application domains (Guerin 2021). This mechanism assigns meaning to "an ambiguous piece of data by leveraging higher level prior knowledge and other available data," which essentially means a given AI application learns on its own without necessary input from the coder(s) or programmer. Geurin brings up the example of hierarchical compositional knowledge structures in which layers (such as a face) describe associations with lower layers (such as an eye or nose). The projection mechanism enables the AI application to create a mapping between elements that it senses from real world data (such as parts of a mouth) and then these are combined with abstract knowledge structures (such as a symbol of a mouth) and the mechanism then defines the structure that can make sense or allow associations between these components through additional recognition of objects, words, and even events (Guerin 2021, 3). This mechanism can also make novel predictions based upon abstract knowledge which then translate into real world elements. If a given AI application includes this projection mechanism, then it can make novel projections because it can pick up one component of say, a face, then search for other components.

Interestingly enough, when projection is implemented in computers, Guerin argues that the selection processes that get implemented through multiple models are stored in "long term memory" and are selected as candidates for interpreting data. This selection process is argued to be beyond our scope of cognition because of their inner-constitution of independent reasoning processes. In the context of robotics, a projection mechanism can help robots to advance in applications in what Guerin refers to as "open environments" comprised of everyday tasks. This will feature the combination of perception and action in that robotic motions will be parameterized by values that they return from perception (Guerin 2021, 11). Guerin foresees that robot skills will enable it to act and learn from behavior and develop new understandings of motor skills. The projection mechanism is key here because, "The application constitutes a sort of reasoning or judgment: if the robot finds a mapping of elements and is confident that it should work, then the robot has made the judgment that this scenario is one belonging to the class where the skill can apply" (Guern 2021, 13). Other mechanisms of AI are numerous and as time goes on, their application will only increase in scope (Anisin 2022a;b).

One particular manifestation of an artistic creation that attempts to depict a robotic future can be observed in the 2014 science fiction movie *Ex Machina*. This movie is about a programmer (Caleb) who ends up falling in love with his humanoid robot in an upscale but remote cabin in the woods. With a 7.7 rating on the International Movie Data Base (IMDB), the film is both a cultural (receiving an Oscar for its visual effects) and commercial success. The premise of the movie is set in a luxury home of the CEO (Nathan) of the company that designed the robot (Ava). Nathan asks the programmer to spend time alone in the house with the robot and if he can identify if Ava is emotionally intelligent and conscious of its behavior. Throughout the course of the movie, Ava is confined to a transparent cell with clear glass barriers and it is through this interactive realm that Caleb establishes a relationship with the robot. Caleb is physically attracted to the robot who has a slim physique and proportionate facial structure. Throughout the course of the first two days, there are several power outages in the large house and one happens when Caleb is meeting with Ava. Ava then blames the power outages on the owner of the house. Ava also starts to dress in specific clothes that make it look more human. Afterwards, Caleb speaks to Nathan who says there is a risk that Ava will get its memory erased which upsets Caleb. Nathan then gets drunk and falls asleep and Caleb steals his login information which he used to enter the room that Ava is in. In the room, he goes through Nathan's computer and identifies several other templates of female-molded robots that Nathan had designed in times past.

He then explores more of Nathan's materials and finds disfigured robot parts laying around in shelves. Caleb encounters Ava and tells it to trigger a power outage so they can escape together. At this point in the story, Caleb's bond with the robot symbolizes a highly irrational form of love that is reminiscent of how love often gets portrayed in film. The following morning, Caleb meets Nathan and the latter reveals that he was secretly spying on the former. Nathan also tells Caleb that he has developed a romantic connection with the robot because it was programmed based on Caleb's past behavior and search terms on the internet (e.g., cookies and data), that Nathan used to calibrate the robot's behavior with. A turning point arises when Nathan tells Caleb that Ava does not really have genuine feelings for him, but that it is simply behaving that way as part of his experiment which he claims has been a success because Caleb fell for Ava. The power in the entire house then goes out and Nathan hits Caleb who falls, then Nathan goes to terminate the robot. Ava retaliates and attacks Nathan who destroys her arm. Nathan dies during the interaction while Caleb makes his way into the room –where Ava tricks him and locks him inside. The robot then goes into the shelves where all other old disfigured robot parts are stored. It finds fake skin and new clothes and dresses herself up to mimic a very human-like organism. Caleb, still locked in the room, can only watch in awe as the robot then leaves the house and goes to the area where Caleb was planned to be picked up from. The robot gets picked up and ventures off into society.

Although just a movie, such robotic creations are not far away and surely will someday become a regular feature of future highly digitalized society. AI is configured by applications which are based on human cognition because programmers write the codes and models that enable it to function, but importantly, AI has the ability to compute its own processes which results in artificial mechanistic behavior and decision making. In other words, AI utilizes finite components drawn from human intellect (through the code that programs a given intelligence machine), but then develops infinite models to influence outcomes. In this sense, it both substitutes and improves upon the analytical aspects of human cognition with computation. The interesting dynamic here is that from a Kantian point of view, AI represents the manifestation of our a-priori knowledge that then coalesces with itself through circuits, microchips, and semiconductors to then eventually undergo a metamorphosis that even Kafka would be surprised by. It took a-priori knowledge and our awareness of space and time to initially produce the foundation of computers then eventually, via software-based machine learning and deep learning, the subsequent manifestation of AI applications led to current systems that can create and write their own language (LaFrance 2017).

As digital innovation progresses into quantum computing, the abilities of intelligent machines will continue to change our understanding of causal forces that bring empirical phenomena about. More significantly, AI based mechanistic actions will arguably lead to some of the most important decisions of human history getting made–such as the decision to launch a retaliatory nuclear weapon during conflict. Social causality is now in part, independent of us, as it controls us via AI and algorithmic manipulation. War reporting, fights on the streets, altercations between civilians, jokes, pranks, love proposals, and nearly all spheres of social life are now warped into stories, images, and videos that if possessing correct (viral-attractive) characteristics, get allocated to top slots by AI, and in turn, this shapes our understanding of empirical phenomena. Perhaps the most significant immediate realm in which AI effects (and will continue to effect) social outcomes and causality in the social world is in the general category of virtual reality.

A Change in How Phenomena are Perceived

AI and its ability to function represents a new evolutionary step in the etymology of the mechanism as there are now mechanistic actions which exist outside of our control, and at the same time, are not simply imagined by us. AI can be considered to be the antithesis to theistic forces that were historically associated with mechanisms for thousands of years. The most significant realm in which AI will manifest is in the social realm –social causality has already started to fundamentally transform because of digital means of communication and what are best considered digital means of living. The aforementioned extent of daily smartphone usage will only increase in scope and capacity as individuals adversely transcend their old way of social organization into digitally based venues. Taking photos of one's body, face, food, interactions, and daily life will become a thing of the past, as newer platforms (e.g., Tiktok) demonstrate how swaths of young people are already starting to move into and adopt platforms that are based on the sharing of video information of one another.

The key issue at hand here is that the trillions of different uploads and exchanges of information and data that get made through digital public realms are not randomly distributed or uploaded and then diffused to populaces in an organic manner. As of 2018. 3.7 billion people used the internet across the world; Google processed over 40,000 searches every second (over 3 billion a day), and every minute of each day saw 46,000 photos get uploaded to Instagram, 456,000 tweets sent on twitter, and over 4 million videos watched on YouTube. What's more, each minute more than 156 million emails are sent as well as over 100 million spam emails. There are also tens of thousands of trips taken via electronic taxi services (Uber) and the weather channel receives an estimated 18 million requests per minute (Marr 2018). Indeed, not all of these services dictate our behavior or cause social outcomes which is why attention must be paid to the specific realm of social media and its associated virtual realities. Here, popularity bias reigns supreme in that recommendation systems which are inbuilt into social media by algorithms are premised on the idea that highly liked items (that get posted, and from their outset, receive a lot of attention) warrant even more recommendations. This results in what computer scientists refer to as complex contagion in which the more times people get exposed to a given idea, story or video, the likelier they are to adopt it and re-share it (Menczer 2021).

The ethical dimensions attached to these processes are quite contradictory. If content that makes it into the "viral" or more most shared categories in a given social media platform, it does not necessarily mean that the content itself is 1) an accurate portrayal of what happened in a given incident or event, and 2) is representative of phenomena that are politically neutral. Beginning with the first of these points, when a video or image is posted on social media networks, it very rarely visualizes the entire setting or context in which it occurred. Videos are most commonly recorded by a given person from a side angle and in rare instances they can be drawn from public cameras. Both of these recording formats and angles do not give the same depth of perception that one's own eyes would if he/she were actually at the scene of an event. Similarly, recordings and photos tend to feature the height or climatic point of a given event or interaction and thus when other people view such phenomena, they do not have a full grasp on the basis of the situation which occurred. In terms of the second point, as more and more people switch over to digital platforms and social media to get informed about socio-political reality (and increasingly, most aspects of reality), the degree of political manipulation populaces are exposed to has reached never-before-seen heights. In the lead up to the 2020 US election, up to 140 million Americans (most of the adult population) received and were exposed to content that had an origin in what are referred to as "troll farms"–which tend to be offshore adversarial entities (e.g., Russia or China) that put out fake content in attempt to stir up political cleavages in a host population. In the US context, these troll farms put out content that pertained to Christian groups as well as African Americans (Hao 2021). In October 2019, Facebook identified 15,000 different pages (which comprise information-generating content that gets sent to users who "like" a given page) and whose audience was a majority American These pages however, were registered by entities in Macedonia and Kosovo. They reached an estimated 75 million users per month. What's more, these "troll farms" have even effectively bypassed platform security and successfully entered monetization advertising programs which are purchased at a given rate by pages and groups to advertise their products. These "instant articles" and "ad breaks" programs in Facebook were utilized by troll farms in such a way that resulted in fake content going unnoticed as it would be mixed in with real content (Hao 2021).

The logics associated with troll farms are essential for our understanding of social causality in the contemporary age (and for its future manifestation) because they entail that content which originates from troll farms and is put out by such entities also enters into the algorithmic makeup of a given social media platform and then, regardless of whether the content itself is actually real, can make it into the "viral" or top trending categories of information that hundreds of millions of people end up seeing. This is a conceptually complex process because it features the production and mental imagery of fake content which then interact with AI-based dynamics. In conjunction, this results in a portion of phenomena and causal outcomes that hundreds of millions (and likely billions) of people perceive are real when they are actually manipulative in their derivation. These are not just sit-at-home processes, but recent history has shown that fake information and content put out by troll-based groups has led to real life actions and events unfolding. Fake information and troll farms have contributed to and even taken advantage of increasing polarization among the US populace by spurring protests and clashes between civilians (Kossof 2017).

The empirical manifestation of such outcomes stem to AI-based algorithmic manipulation and importantly, such outcomes do not arise simply due to the existence of AI, but are deeply rooted in what can be considered nothing other than the "hacking" of human beings' dopamine system and reward structures. Dopamine is a chemical and hormone that gets produced by our brain naturally in response to stimuli whether it be food, exercise, drugs, sexual activity, positive reactions, among many other different phenomena. There are four primary dopamine pathways that scholars have identified and each has its own set of neurotransmitters that bring about cognitive and motor movement outcomes based on rewards. In individuals that have suffered from addictions, these pathways are dysfunctional. Positive stimuli that stem to the social world result in the brain releasing dopamine which becomes a reinforcive process based on the antecedent stimuli that caused the release. Krach, Paulaus, Bodden, and Kircher (2010) demonstrated that positive recognition by peers activates dopaminergic pathways which is why social media has been considered by many to be an addictive activity–with every like, retweet, or comment, the attention a given person receives triggers his/her dopamine pathways, which in turn, rewires the brain into a new addictive configuration that thrives on such behavior. Haynes identifies the adverse nature of these processes by correctly pointing out that,

> "Because most social media platforms are free, they rely on revenue from advertisers to make a profit. This system works for everyone involved at first glance, but it has created an arms race for your attention and time. Ultimately, the winners of this arms race will be those who best use their product to exploit the features of the brain's reward systems" (Haynes 2018)

What's more, Haynes (2018) argues that social media applications have been particularly designed to "take advantage" of our dopamine pathways to literally hijack our attention and make our usage of such platforms increase into something that is habitual, rather than voluntary. This is precisely why if one goes to any given metropolitan area in the world and rides on public transport for at least an hour, he/she will notice that a majority of people are on their smartphones. People are not just scrolling through their contacts or sending messages to others, but are most commonly scrolling through social media platforms and watching content that is provided to them via algorithmic manipulations featuring real information, fake information, and above all, information that is meant to gain the most attention possible and prolong a given user's engagement on the platform. Moreover, these processes have become so immoderate that algorithms are inbuilt into social media platforms to specially withhold likes from a user only for the purpose of delivering them in larger bursts when they pile up–resulting in an initial downturn (e.g., a person posts something but does not receive the amount of likes he/she desires), but then after some time, a heap of likes and comments follow to reward the user with a greater dopamine burst (Haynes 2018).

This brings us back to a salient development that Berdyaev accurately predicted. He foresaw that this terribly augmented power developed by human beings over nature would lead to our enslavement to our own discoveries (Berdyaev 1952, 47). Is not the current landscape of smartphone usage in which most adults (and teenagers) spend half of the entire day on their phones not an enslavement to our own discoveries? Are not the algorithmic biases and manipulational pathways associated with using social media platforms a concealed form of subjugation that adversely construes our knowledge of the world? These processes will only intensify as time goes on because with each young generation that emerges, more time gets spent by young people on social media and digital platforms. The generation known as "z" is particularly interesting because it has already provided us with empirical evidence that can lead one to postulate what future generations' behavior will be like. Generation z is the first generation in all of human history that has grown up within the internet. Instead of relying on close social bonds through in-person interactions, many from this generation have relied on google searches, forums (e.g., Reddit or Quora), and other platforms to seek out information on how to behave and make decisions. In my view and experiences with hundreds of students that make up this generation, their behavior is marked by a huge drop off in emotional intelligence, increased shyness, and anxiety which are observable throughout interactions that take place in classroom settings. The markedly different behavior of this generation stems to the fact that large portions of it have grown up alongside smart devices and digital platforms, and have spent less total time outside interacting with peers and other members of society than generations of the past. It may be the case that Dreyfus' arguments were wrong, as he simply could not foresee that human beings would end up growing up largely through technology, and in that sense, future generations would become more like AI. Moreover, recent survey research put out by the American Psychological Association has indicated that members of generation z experience more stress on average than adults, and especially with relation to news about societal problems. Generation z report greater levels of problems in their mental health and on average (more than 9 in 10 of generation z participants in a survey indicated they have experienced at least one adverse physical or emotional symptom because of stress). This generation has already received a higher percentage of treatment for their mental health than all other current generations that are alive (Bethune 2019).

I recall my own time as a student and early years taking college courses–short breaks that would be taken in the middle of classes were times of conversation and reflection. Barely anyone had their mobile phone with them in break. During these breaks, some went outside, to the bathroom, to smoke, to get a coffee, while most remained in class and engaged in organic conversations about a variety of topics. Many friends were made during these breaks and connections were established that carried on over to other settings when I would encounter people from my classes–as students, we had established real connections and experienced a similar reality that was in no way influenced by exogenous, algorithmic forces. Everything was on the table, so to speak, and it was through our very similar experiences that we were able to help each other learn and cope with challenges that arise in life during early adulthood. After weekends, we all asked each other what we did and fun stories were told. Today, most young people know what all their peers are doing, and fiendishly follow their updates on an hourly basis on social media platforms. To contrast these experiences with what happens in contemporary classrooms in the digital age–during breaks, 90% of students that remain in the class pull out their smartphones and sit on their devices the entire break; it is very rare that any conversation starts at all. These students are only partially present in their current reality because they are fully immersed into their virtual realities–every chance they get, they look at their smartphones and check for notifications and other messages. It is thus no surprise that generation z and even the millennial generation (which also spends a great deal of time on digital platforms and the internet), experience the greatest problems in mental health that arguably have ever been observed by psychiatric specialists. A study put out by the Harvard Business Review revealed findings that are nothing short of shocking – half of millennials (age 24-39) said they had left a job “at least partly for mental health reasons,” while 75% of generation z (age 18-23) left a job for such a reason. These are astronomically high numbers compared to the rest of the general population of which only 20% reported such adversities (McMaster 2020). Likewise, in 2020 at the University of Alberta (Canada), 35% of the entire student body reported that they experienced a panic attack due to stress (McMaster 2020), which indicates there is something fundamentally augmented and harmful built-in to current communication and social behavior patterns.

The advent and subsequent success of Virtual reality (VR) devices has led to the formation of fabricated and fictitious worlds in which individuals can immerse themselves in. VR is defined as a combination of hardware and software systems that seek to create a perfect environment that is all-inclusive of sensory illusions (Radianti et al. 2020). At the time of writing this book, VR is typically experienced by devices that one wears on his/her head, covering the eyes and is often accompanied by headphones. Scholarly literature on VR is booming and much has been said thus far about the application of VR to educational and therapeutic fields (Radianti et al. 2020). Just as many other historical processes in which technological developments arose, entire societies and populaces simply gravitated towards their usage with little to no critical reflection. Drawing back to Heidegger, the implementation of VR in educational settings signifies how humans are simply immersed into a large-scale technological process that treats them as units which need optimization. The implications of this ontological condition is that classical forms of learning (such as reading physical books, writing on paper, chalk/white boards, etc), are automatically replaced by new technological gadgetry simply because the advent of new technologies is assumed by either technocrats, public policy makers, or large portions of publics to be better than older traditions of learning. With each new technology that arises, it will override and attempt to dominate preceding methodologies of human organization and throughout this course of changes, individuals and individual experiences will become de-humanized in favor of virtual and alternative realities.

These processes are already taking shape and having real world impacts on macro-level social outcomes. The inertia associated with reliance on digital platforms and content to live one's life, interact, and obtain information has taken an adverse toll on the development of young people. Young people have failed to embrace adulthood, and we are currently witnessing historically low levels of young people obtaining drivers licenses–87% of 16-year-olds had drivers licenses in 1983 while only 25% did in 2014 (McMaster 2020). Why drive and explore the world if you can get goods and food delivered to your door through a smartphone app? Why take initiative and learn something new when you can drive in a video game? These are questions that are parasitic on the post-modern condition. All in all, both Berdyaev and Heidegger seem to have predicted and accurately summarized an era whose foundation was designed and engineered during the end of their lifetimes, then developed in and came into fruition decades after. Berdyaev saw great trouble brewing for the future in that our relationship with technology would end up enslaving us to our own discoveries, while Heidegger similarly believed that human beings will become de-humanized and simply morph into being a part of technological processes–as units and materials that need to be optimized. It is feasible to believe that what we are currently witnessing in the high daily usage of smartphones and digital platforms will only increase in scope and the aforementioned adverse outcomes associated with these tendencies will only aggravate.

As Berdyaev noted in one of his last written works, "Machines are made from material elements taken from the old nature, but into them goes something quite new, no longer of nature, and not a part of the old cosmic order" (Berdyaev 1952, 51). More than two millennia ago, the idea of a mechanistic force was portrayed through a theatrical design in Ancient Greece. This contributed to narratives of how gods and goddesses determined outcomes through the primary mode of entertainment and storytelling of that era–theater. Long after, the digital revolution arose and here the mechanism was redesigned into a circuited, micro-chipped, intelligence system that runs on programming code–code which specifies the exact function of a given AI system through algorithms and utility functions. These algorithms and mathematical axioms determine the things billions of people see on a daily basis and form new causal processes that will continue to manifest independent of our mind.

References

Adams, Tim. 2016. "Artificial Intelligence: 'We're Like Children Playing With a Bomb.'" *The Guardian*, June 12, 2016. https://www.theguardian.com/technology/2016/jun/12/nick-bostrom-artificial-intelligence-machine

Anisin, Alexei. 2022a. "Heidegger and the Technocratic Warping of the COVID-19 Pandemic." *Cultural Studies < = > Critical Methodologies*. https://doi.org/10.1177/15327086221087651

Anisin, Alexei. 2022b "Pandemic Surveillance Capitalism: Authoritarian Liberalism or Democratic Backsliding?" *Journal of Political Power*. 1-17. https://doi.org/10.1080/2158379X.2022.2055275

Berdyaev, Nikolai. 1952. *The Realm of Spirit and the Realm of Caesar*. London: Donald A. Lowrie.

Bethune, Sophie. 2019. "Gen Z More Likely to Report Mental Health Concerns." January 2019. American Psychological Association, *Monitor on Psychology*. https://www.apa.org/monitor/2019/01/gen-z

Bostrom, Nick. 2014. *Superintelligence: Paths, Dangers, Strategies*. Oxford: Oxford University Press.

Ceci, Lucie. 2022. "How Much Time On Average Do You Spend On Your Phone On a Daily Basis?" *Statista*, February 2022. https://www.statista.com/statistics/1224510/time-spent-per-day-on-smartphone-us/

Dreyfus, Herbert. 1972. *What Computers Still Can't Do: A Critique of Artificial Reason*. Cambridge: Massachusetts Institute of Technology Press.

Dreyfus, Hubert. 1995. "Heidegger on Gaining a Free Relation to Technology." in *Technology and the Politics of Knowledge*, edited by A. Feenberg and A. Hannay. Bloomington and Indianapolis: Indiana University Press.

Fakhrutdinov, Rafael. 2018. "Doomsday Machine: Poseidon." *Gazeta.ru,* December 29, 2018. https://www.gazeta.ru/army/2018/12/29/12113839.shtml

Ferris, Robert. 2017. "Elon Musk Thinks We Will Have to Use AI This Way to Avoid a Catastrophic Future." *CNBC*, January 31, 2017. https://www.cnbc.com/2017/01/31/elon-musk-thinks-we-will-have-to-use-ai-this-way-to-avoid-a-catastrophic-future.html

Guerin, Frank. 2021. "Projection: A Mechanism for Human-like Reasoning in Artificial Intelligence." *arXiv* preprint. https://doi.org/10.48550/arXiv.2103.13512

Halpin, Harry. 2004. The Semantic Web: The Origins of Artificial Intelligence Redux. In, *Third international workshop on the history and philosophy of logic, mathematics, and computation* (HPLMC-04 2005). San Sebastian, Spain.

Hao, Karen. 2021. Troll farms reached 140 million Americans a month on Facebook before 2020 election, internal report shows. September 16, 2021. Technology Review. Retrieved from: https://www.technologyreview.com/2021/09/16/1035851/facebook-troll-farms-report-us-2020-election/

Haynes, Trevor. 2018. "Dopamine, Smartphones & You: A Battle For Your Time." Harvard University, *The Graduate School of Arts and Sciences*. May 1, 2018. https://sitn.hms.harvard.edu/flash/2018/dopamine-smartphones-battle-time/

Heidegger, Martin. 1977. "The Question Concerning Technology," in *Martin Heidgger: Basic Writings*. Edited by David Farell Krell. New York: Harper & Row.

Kelleher, John D. 2019. *Deep Learning*. Cambridge, MA: Massachusetts Institute of Technology Press.

Kossof, Maya. 2017. “The Russian Troll Farm that Weaponized Facebook had American Boots on the Ground.” *Vanity Fair*, October 18, 2017. https://www.vanityfair.com/news/2017/10/the-russian-troll-farm-that-weaponized-facebook-had-american-boots-on-the-ground

Kurzweil, Ray. 2005. *The Singularity Is Near: When Humans Transcend Biology*. London: Penguin Books.

Krach, Sören, Frieder M. Paulus, Maren Bodden, and Tilo Kircher. 2010. "The Rewarding Nature of Social Interactions." *Frontiers in Behavioral Neuroscience* 4 (22).

LaFrance, Adrienne. 2017. An Artificial Intelligence Developed Its Own Non-Human Language. *The Atlantic*, June 15, 2017. https://www.theatlantic.com/technology/archive/2017/06/artificial-intelligence-develops-its-own-non-human-language/530436/

Lyotard, Jean-Francois. 1984. *The Postmodern Condition: A Report on Knowledge. Vol. 10*. Minneapolis: University of Minnesota Press.

Marr. Bernard. 2018. “How Much Data Do We Create Every Day? The Mind-Blowing Stats Everyone Should Read.” *Forbes*, May 21, 2018. https://www.forbes.com/sites/bernardmarr/2018/05/21/how-much-data-do-we-create-every-day-the-mind-blowing-stats-everyone-should-read/?sh=359e856b60ba

McCarthy, John. 1989. "Artificial Intelligence, Logic and Formalizing Common Sense." In *Philosophical Logic and Artificial Intelligence*. Edited by Richard H. Thomason, 161-190. Dordrecht: Springer.

Menczer, Filippo. 2021. “Here’s Exactly How Social Media Algorithms Can Manipulate You.” *Big Think*, October, 7, 2021. https://bigthink.com/the-present/social-media-algorithms-manipulate-you/

Nahin, Paul. J. 2017. *The Logician and the Engineer: How George Boole and Claude Shannon Created the Information Age*. Princeton: Princeton University Press.

Neuralink. 2022. “Approach.” *Neuralink*, March 2022. https://neuralink.com/approach/

Nilsson, Nills. 2010. *The Quest for Artificial Intelligence: A History Of Ideas and Achievements*. Cambridge: Cambridge University Press.

McMaster, Geoff. 2020. “Millennials and Gen Z Are More Anxious Than Previous Generations: Here’s Why.” *University of Alberta*. January 28, 2020. https://www.ualberta.ca/folio/2020/01/millennials-and-gen-z-are-more-anxious-than-previous-generations-heres-why.html

Radianti, Jaziar, Tim A. Majchrzak, Jennifer Fromm, and Isabell Wohlgenannt. 2020. "A Systematic Review of Immersive Virtual Reality Applications for Higher Education: Design Elements, Lessons

Learned, and Research Agenda." *Computers & Education* 147: 103778.

Regan, Natasha and Matthew Sadler. 2019. DeepMind's Superhuman AI is Rewriting How We Play Chess. *WiredUK*, February 3, 2019. https://www.wired.co.uk/article/deepmind-ai-chess

Riehm, Kira E., Kenneth A. Feder, Kayla N. Tormohlen, Rosa M. Crum, Andrea S. Young, Kerry M. Green, Lauren R. Pacek, Lareina N. La Flair, and Ramin Mojtabai. 'Associations Between Time Spent Using Social Media and Internalizing and Externalizing Problems Among US Youth.' JAMA Psychiatry 76 (12): 1266–1273.

Ruttan, Vernon. 2006. *Is War Necessary for Economic Growth?: Military Procurement and Technology Development*. Oxford: Oxford University Press.

Vice News. 2019. "The Race For Quantum Supremacy." *VICE on HBO*, October 30, 2019. https://www.youtube.com/watch?v=1lIfbqfoGMo&t=190s

Weinberger, Sharon. 2014. "Why Google and The Pentagon Want 'Quantum Computers." *BBC Future*, November 18, 2014. https://www.bbc.com/future/article/20130516-big-bets-on-quantum-computers

Wilks, Yorick. 2019. *Artificial Intelligence: Modern Magic Or Dangerous Future?*. London: Icon Books.

Wooldridge, Michael. 2021. *A Brief History of Artificial Intelligence: What It Is, Where We Are, and Where We Are Going*. New York: Flatiron Books.

CHAPTER SEVEN

CONCLUSION

As the ink in a writer's pen dwindles down, the tip reaches a dry point and makes a scratching sound which indicates a book is closer to the end rather than the beginning. With any work whether it be non-fiction or fiction, the concluding segments of a book always feature a degree of reflection and process of pondering where an author asks whether the preceding chapters fulfilled the aims that were set out when beginning the book. In my case, the steps taken while doing research and engaging in theorization while writing this work led to engaging and unexpected terrains–these terrains lead me to conclude that the understanding that human populations have had about social causality has historically varied from the onset of the creation of metaphysics in Ancient Greece to the emergence of the era of Christianity, into the middle ages, the enlightenment, the era of scientific experimentation, and into our contemporary era of digitalization. Likewise, the ideas associated with causality in the social world varied and transformed throughout all of these different eras of human history. This book has mapped out and identified a heterogeneous collection of changes that were associated with the semantic characteristics of the term "mechanism." This term is obviously not comprehensive of all phenomena associated with social causality, but it is arguably the most important and consistent term that has been implemented in cross-historical settings to explain causes in different contexts and circumstances. I hope that this book has offered both students and scholars a new angle with which to read into and explore age-old debates found in philosophy, the philosophy of science, and the philosophy of social science.

In the second chapter, an overview of the intricate historical process of the development of the natural and social sciences was provided which began with some of the first creations that humans used to improve their chances of survival. Although early civilizations such as Ancient Egypt or Mesopotamia did exert a certain degree of control over nature, it was not until the Ancient Greeks that broad ideas on experimentation, logic, and metaphysics emerged. Both Plato and Aristotle's contributions were given attention as well as noteworthy figures of the time including the Pythagoreans.

The chapter then delved into the different discoveries that were made in the "dark ages" as well as in the subsequent Middle Ages–with emphasis placed on the Greco-Roman world and especially the Western European context. The chapter then over-viewed changes in society that arose by the 17th century which resulted in the mechanization of nature and concurrent rise of modern science and experimentation. The philosophical contributions of Kant were given attention as well as Hume. The chapter also identified the rise of social science in the 19th century in contexts that experienced rapid industrialization. The book's first distinction between natural and social science was provided which paved the way for direct assessment of modern conceptions of mechanisms in subsequent chapters.

In the third chapter, I identified the origin of the term mechanism (mechane) in the Ancient Greek context and delved into the underpinnings of an era that featured the first formulation of what later turned out be the field of philosophy. Even with the advent of logic and a preliminary degree of scientific reasoning, this era was marked by polytheistic belief systems and populations attributed major social outcomes to theistic forces. This is why the deus ex machina (god from the machine) was the most symbolic and powerful manifestation of social causality–being witnessed by tens of thousands in theater settings that saw the portrayal of gods and god-like forces in some of history's greatest tragedies. Several noteworthy theatrical presentations were covered including Euripides' *Medea*, *Orestes*, and Aristophanes' *Peace*, all of which shed light on the general spirit of the polytheistic belief system in which the idea of a mechanism originated from. This chapter demonstrated how early metaphysical conceptions of causality led to the creation of the mechane and the widespread usage of a theatrical tool. This tool was supported by a feat in mechanical engineering which exemplified metaphysical notions of cause and effect that were based around divine intervention in theatrical tragedies. The fascinating findings of this chapter (and these findings were unknown to the author before carrying out research on Ancient Greek theater) is that the deus ex machina that emerged in this context was not only the first conception of a divine and metaphysical force that dictated plot shifts and story endings with regularity in a given socio-cultural context, but that actual theatrical shows were complimented by the mechanical engineering of mechanes that would enable this concept to be implemented for audiences. This exemplifies human beings' a-priori abilities to judge the world according to space and time–engineering a mechane was a sophisticated endeavor and required knowledge of correct parameters of length, weight, and material. These cognitive resources were used to then exemplify widespread societal beliefs about the metaphysical characteristics of historical change and daily life. The chapter also connects the idea of a deus ex machina to social causality in the Ancient Greek context and makes the case that subsequent semantic changes that arose with this term reflected changes in ontological understandings of social causality.

The fourth chapter delved into the Middle Ages–a vast time period of history that was marked by a relatively stagnant Europe and a flourishing China. Of all chapters in the book, this one was the widest in terms of the sheer time period under attention which also pertains to the relative staleness that European civilization experienced throughout a span of more than one thousand years. This time period was marked by the fall of the Roman empire and the rise of Christianity. The latter was especially important for how previously salient polytheistic beliefs associated with social causality gradually receded and was followed by the rise of a single theistic force that was representative of one God. Early Christian writers such as St. Ignatius of Antioch articulated ideas about how the idea of human beings' connection to God could function through a cosmic hoist that would enable people to reach heaven. With the decline and abashment of paganism, two strains of Christianity ended up reigning supreme in Western Europe (Catholicism) and Eastern Europe (Orthodoxy). This profound shift in belief systems had significant implications for how populations interpreted social outcomes. Most populations in across the European continent lived in settings that were wholly dictated by Christian faith and forms of social organization were heavily embedded within Christian customs. The chapter illustrates some of the tendencies that arose and were associated with the new individuation of the deus ex machina in classical literature through the stories of the *Tale of Gamelyn*, Shakespeare's *As you Like it* and *Hamlet*.

In the fifth chapter a comprehensive account of the rise of science was provided along with the conceptual distinctions that philosophers of science have put forward on the usage of mechanistic explanations. The chapter also placed great emphasis on the fact that with the rise of modern science, populations gradually shifted from highly theistic to atheistic settings of social organization and this had a major impact on the autonomy that specific groups and individuals could experience with relation to identifying laws, regularities, and mechanisms. For the first time in human history, comprehensive attempts were launched at identifying tendencies in the natural world that were not assumed to be brought about by anything other than material and natural phenomena. The chapter identified a key characteristic that is inherent to mechanistic explanations in the natural sciences through the Krebs cycle and associated causal processes that are driven by mechanisms. Mechanisms in the natural world are much more stable and consistent across space and time than they are in the social world. Examples to support these claims were also put forward through illustration of mechanisms that link obesity and cardiovascular disease. The chapter then shifted to accounting for the rise of mechanistic explanations in the social sciences – which was demonstrated to constitute a much newer form of explanatory trend and one that is marked by substantial inconsistency with relation to how mechanisms are defined and applied in empirical analyses. The chapter concluded with two examples drawn from own research endeavors in which I attempted to account for both qualitative and comparative dynamics and mechanisms inherent to the phenomenon of repression backfire from the state repression and nonviolent civil resistance literatures.

The sixth chapter was the final of this book's direct analysis of semantic changes of the mechanism and its relationship to social causality. The chapter accounted for the rise of modern technology through drawing on both Berdyaev's and Heidegger's ideas on macro historical societal processes. In essence, both philosophers made indirect predictions about what they believed the future of a highly technological society would comprise and both turned out to accurately account for such developments. The chapter also explained the origin of the internet through emphasizing its links to government defense projects and funding. It then identified the basis of artificial intelligence and explored its application throughout different spheres of life and social organization. The chapter put forward a provocative argument by stating that the contemporary age of digital internet connectivity and communication is the first in history to feature social outcomes and causality that are brought about by forces that are independent of our mind. Even though AI applications are coded and created by human beings, their empirical application can (and will increasingly) surpass the cognitive abilities of the creators of such applications. The chapter demonstrated that such a process is already presenting grave ramifications for populations who are unaware of the nature of the digital platforms that they spend more than half of their day on which brings us to the concluding findings of this book.

Point One

There are five major points that readers can take away from this book. First, semantic changes of the term mechanism have been associated with grander shifts in views of social causality across different historical eras. This begs the question, so what then, is a mechanism? This book has demonstrated that this term reflects societal views about causality. At first, the term was conceived through populations' attempts to rationalize their beliefs about the nouemenal world into the phenomenal (e.g., the Ancient Greeks). This resulted in polytheistic-derived views of social causality, which then morphed into theistic views with the rise of Christianity and the decline of paganism. Such views were prominent until the emergence of the era of modernity and the rise of science, which resulted in notions of social causality being associated with the noumenal world, and in turn, the mechanism was interpreted as an empirical regularity that was not connected to any theistic force or power. In the final stages of this metamorphosis, mechanisms of artificial intelligence have started shaping social causality and this trend is unlikely to slow down.

Through exploring different eras and time periods of human history, this book has argued that the term mechanism is reflective of a larger understanding or set of understandings that pertain to commonly held worldviews by a population. Such an analysis would not have been possible to put forward or initially be conceived of if there were not both regularities and differences in worldviews across space and time. At the outset of writing this book, it was clear that earlier societies and civilizations held beliefs about social causality, and that these beliefs were expressed with regularity in those societies. Yet as eras of history changed, as dozens of new generations arose, and as empires crumbled and new ones rose, innovative views about the world appeared. Across time, there has been a lack of uniformity in worldviews which is fascinating from both philosophical and social scientific perspectives as this variance provides us with the opportunity to investigate both why and how changes occurred and what the implications of these changes are for the semantic characteristics associate with the term mechanism.

The mechanism is a force that has widely been conceived as causal but importantly, its conceptions have led to the creation of something which is physical (i.e. the literal creation of a mechane in Ancient Greek and Roman theater settings), metaphysically abstract (the deus ex machina as a plot shifting mechanism), real (such as the interactions of molecules in a biological process), empirically abstract (such as the classification of different components of social behavior including antecedent conditions in a process model and mechanistic outcomes such as emotional outrage), and finally, as something that is both concrete and abstract (mechanisms in artificially intelligent machines and applications). The mechanism has reflected what populations believed to constitute the ultimate causes associated with occurrences, events, and manifestations of phenomena in a given time period or era of history. In Ancient Greece, these worldviews were profoundly polytheistic and driven by beliefs in heterogeneous gods and phenomena associated with gods. The creation of the mechane in Ancient Greece resembles what Heidegger described as a process of bringing forth (poesis)–this was a highly creative endeavor that was not necessarily done to accomplish a means to an end set of tasks. In the subsequent era of Christianity, this heterogeneity slowly morphed into homogeneity–with Jesus Christ and the image of the cross leading to widespread changes in worldviews that constituted a unitary function–a function that saw most of social organization and life across Europe being driven by belief in one God and the manifestation of many different societal customs associated with the New Testament. This dominant belief system started to rapidly decline in its societal influence by the time of modern history, industrialization, and the rise of scientific experimentation. Mechanisms became synonymous with only science and scientific explanations (both natural and eventually social science) because of the profound ability that humans developed to change the world, master control over natural phenomena, and to extend life based on scientific discoveries such as heating, electricity, bacteria, vaccines, etc. Into the forthcoming era the digital age arose and compared with all eras of the past, outcomes in the digital age are continuously brought about by mechanistic forces which are underpinned and driven by autonomous spheres of artificial intelligent computer applications.

Retrospectively, we should not be surprised that the meaning of what constitutes a mechanism has undergone changes and has transformed in accordance with societal processes and changes in worldview–the shift from polytheism to theism was profound in its manifestation in social organization across Europe–a process that unfolded over the course of more than 500 years. No less profound was the mass alteration that arose along with modern society and science. Perhaps the most profound change, however, was in the rise of digital communication and the internet. While the advent of science changed the relationship that human beings had with nature, fostered never-before-seen material wealth, and enabled the average life span to gradually increase in length, the rise of digital technology led to a much more radical set of changes. The internet not only enabled the proliferation of computers and smart devices into being common household items, but it also led to the development of artificial intelligence and associated processes such as machine learning and deep learning– which over the course of just a few short decades have fundamentally transformed the way that human beings behave, perceive social reality, and communicate with others.

Indeed, differences in urbanization between the period of history marked as the modern era were profound in comparison to the Middle Ages where most people in Europe lived in rural settings and relied on agricultural production to survive; yet into the modern era, the medium of communication that people relied on was relatively static–people spoke to one another in person, attended meetings, and for those that could read, they read paper-based information. Newspapers, telegraphs, and eventually telephone and radio did improve communication capabilities between members of a given populace, but they did not present a digital realm in which human beings would spend a majority or more of their day using. Since the rise of digital technology has already resulted in the solidification of a computational realm–this has forged a set of social relations that differ from anything that has been experienced or created in the whole of recorded history. The profound impact that the digital age has had on social causality cannot be overstated. For the first time in history, information is shared at rapid speed and billions of people spend most of their time strapped to smart devices, starring at screens that spill more content into the human mind than any single science fiction author from the early 20th century could have dreamed of.

Point Two

The second major takeaway point of this book is rooted in an interesting, albeit complicated dynamic related to the contingency and social causality. As the title of this book indicates, social causality has been contingent over the course of thousands of years, and different civilizations have made different contributions to philosophy, science, and to technology. Considering the final or most recent era of digital communication and artificially intelligent computation, can one attribute the term "evolution" to semantic changes associated with the term mechanism since the time of Ancient Greece, into the Middle Ages, then to modern science, and finally into contemporary society? It is clear that the technical capabilities that have arisen over the last two millennia far surpass anything that people of ancient societies could have foreseen or even predicted, but when it comes to assigning an evolutionary classification to the mechanism and its relation to how social causality is interpreted, one may run into difficulties. There surely has been an evolution in our ability to measure the empirical world, manipulate it, and to some degree, exert our control over it. Our control over the phenomenal world has evolved, but it seems that our experience and involvement with the noumenal world has not.

Along similar lines, it is unclear if actual societal understandings of the social causes of outcomes have evolved. Societies have not been able to keep up with the philosophical, scientific, and technical innovations of their own time. Throughout what many believe to be evolution, it seems as if a certain group or sub-set of every society and populace to date have understood the nature of social causality, while masses would then adhere to and follow hegemonic understandings. As such, hegemonic political formations and discourses surely have played a principal role in driving understandings of social causality. In turn, they have contributed greatly to the historical morphing of the mechanism across different time periods.

The analyses carried out in this book have led me to believe that while technical and scientific progress have been stagnant in certain time periods and rampant in other time periods, the greater whole of society throughout times of decay and progress have too often fallen victim to scientific discoveries and have likewise found themselves to be positioned outside of processes. This leads me to conclude that there surely has been technical evolution and the semantic changes associated with the term mechanism have reflected this technical evolution, but societies have not kept up with this evolution on a social level, even if they have been positioned within the greater structure of technical and scientific development. In our current era, the widespread immersion of populations into digital reality and electronic forms of communication has coincided with historically high levels of atheism. Meanwhile, social media and digital forms of communication have brought about a new form of polytheism that can be considered to be a digital paganism.

This new phenomenon of digital paganism is premised on the idea that engagement with social media, public posts of personal information, and immersion into digital realms is a positive activity for individuals to engage in, and within digital realms, "influencers" as well as celebrities and other internet personalities have taken on statuses that are not at all different from polytheistic gods. For example, in 2021, a new trend arose due to an influential TikTok account called "Devious licks" who promoted vandalism and encouraged young students in high schools and middle schools to launch pogroms on their campuses, theft of school items, vandalize buildings, break furniture, trash bathroom supplies, etc. This was referred to as an "October challenge and was not a unique occurrence – other challenges have also arisen which included the encouraging of the purposeful contracting of COVID and even shaving one's teeth with a nail filer. These trends are ritualistic and the influencers that encourage them serve a God-like function for their worshipers in groups and communities in digital platforms. Such examples are not meant to demean contemporary social groups. If we take any period of human history, we will find heinous practices such as religious abuses going on in British churches wherein priests would frequently take sexual advantage over women in confession or the mass amount of violence that was carried out during the inquisition.

Such adverse outcomes pertain to the general point that masses that make up a populace in a given historical time period have tended to adopt understandings of social causality which are infantile compared to the actual processes of technical and scientific evolution that small groups of highly educated members of society understand. The latter processes have served as the driving forces behind semantic changes associated with the mechanism.

Point Three

A third takeaway point to consider is that in the context of the social sciences and humanities, a great amount of ink has been spilled in debates about social construction. There have been immensely antagonistic schools of thought that have argued and debated about whether social reality is socially constructed and whether the regularities and mechanisms that are privileged in positivist modes of inquiry constitute real phenomena or are just scholarly constructions. These debates have gone as far as to say that principal classifications and categories of understanding that have been around for thousands of years (such as gender; male/female) are invalid. This book's analysis of semantic changes associated with the mechanism and with social causality have put forward evidence to support the contingent nature of the term mechanism, but even with this contingency, one must not lose sight of the fact that if phenomena are socially constructed, the human mind's ability to engage in cognition and construction has led to the identification of truthful phenomena and outcomes in the empirical world that are stable across time and space. This is especially salient with reference to the natural world, and a less salient with reference to the social world. In other words, yes, our assumptions of social reality are indeed contingent if viewed from a historical lens and are of course constructed, but this does not mean construction is in anyway wrong or unfruitful. Part of the reason that anti-positivist schools of thought have argued against the utility of rationalist thought and formalistic approaches in the social sciences likely has to do with their lack of understanding of the nature of social causation and particularly the lack of causal uniformity in the social world. The social world is incredibly complex, and when attempting to observe its complexity, it seems that social scientists too often fall into camps (either the constructivist or positivist camp) and either accept mechanistic explanations or reject them on the grounds of the idea that mechanisms are parasitic upon the practices and contexts in which they operate in. Yet this is the precise point–that mechanisms are parasitic upon some contextual conditions in some types of phenomena, yet they also can be general and active across different contexts in other kinds of phenomena and outcomes. If schools of thought do not believe in the utility of social construction, then they simply fall into a paradigm of nihilism that rejects the existence of basic entities of social order and human organization. In this respect, the age-old constructivist versus positivist debates found across social scientific disciplines are better off considered as relics of disciplinary knowledge.

Point Four

This leads us to the fourth and final takeaway point of this book which has to do with Kant's contributions to philosophy. In Kantian terms, the historical creation of a mechane was an attempt to represent the principal source of understanding of power and causality that people held of the noumenal world. In order to mentally forge the mechane, to conceive of its design before its actual physical creation, a-priori knowledge of the phenomenal world was drawn on. The creation of the mechane can thus be considered to represent a partial attempt by the Ancient Greeks to recognize what Kant believed to be unknowable–gods and supernatural forces, both of which belong to the noumenal world. From the outset of writing this book, I never wanted to fully adhere to or utilize one particular philosophical framework or rely on the contributions inherent to one particular strain of thought because of the sheer breadth of the time periods under attention. There have been helpful frameworks drawn upon to account for the historical emergence of phenomena such as technology, modern science, and mechanistic applications in both the natural and social sciences. Yet while writing the concluding chapter of this book, it became clear to me that of the varying frameworks that were drawn on, the Kantian classification of the phenomenal and noumenal worlds appears to be the most relevant.

Recall that the phenomenal world consists of a world that we construct with our empirical senses–we are aware of this world and it appears to be present to us. The characteristics Kant assigned to the noumenal world strongly contrast as he argued this world exists but is unobservable as it lays underneath our empirical perceptions and senses. In multiple ways, these categories reflect my overall summary of the causes of semantic change associated with the mechanism described in paragraphs above, especially with regard to the ideas of social construction and social causality. Specifically, Kant believed that empiricists were correct in asserting that knowledge depends on our empirical senses and sensations, while he also believed rationalists were correct in believing that there are laws which exist in the world (e.g., that outcomes and events have a cause). Since the human mind assigns its synthetic a-priori categories to the world, it is thus no surprise that there have been variant interpretations of causality and mechanistic forces across different historical eras–the semantic changes associated with the term mechanism reflect different individual and collective attempts to rationalize the noumenal world into the phenomenal. Kant believed that reason and rationalization could never address the basis of God or theism because our empirical senses only present us a world that we experience, and may not actually reveal the genuine character of the world–the unknowable. Different societies have thus sought to believe (via free will and other similar concepts) in higher entities for the grander purpose of establishing the possibility of forming moral views.

Importantly, if we consider that Kant also argued that both time and space are not independent of our minds, then it can become precisely clear as to why there have been different conceptions of social causality and mechanistic forces across different eras. Considering that the categories of our mind cannot adequately grasp metaphysical issues including theism (God) and associated themes such as immortality, they also could not disprove the existence of such entities, argued Kant. This is a key aspect in Kant's thought because it entails that there is a certain ontological tension that has underpinned attempts at rationalizing both the empirical world and the noumenal world–in early periods of history, populations attempted to rationalize the empirical world through placing great emphasis on theistic powers and forces, then as time went on and theism declined, more emphasis started being placed on rationalizing the empirical world without any associated theistic caricatures. This led to great scientific discoveries, the identification of mechanistic forces in both the natural world and within our own bodies. Yet the different processes that go on within our bodies (as outlined in earlier chapters of this book), such as the mechanisms of the Krebs cycle or the mechanisms that link obesity to cardiovascular disease, are based upon categorizations that are indeed helpful, but nevertheless fall short of enabling us to accurately predict (with certainty) across populations (e.g., when someone will actually have a cardiovascular event due to their obesity).

Here we arrive at an impasse that Kant obviously could never address (and to the best of my knowledge, conceive of) because of the time period which he lived in was far removed from the era of digital technology and computation. Since Kant argued that objectivity in the empirical world does not exist independent of our mind (i.e., our mind creates the objective world), this fundamental aspect of his framework of categories falls short in being able to account for AI and its emergence. This book has demonstrated (in chapter six in particular) that the rise of AI and its infusion into digital technologies has resulted in something that is of a different technological essence that constitutes processes of social causality which arise outside of our mind's perception of the phenomenal world. Put differently, the ability of AI-based programs to create actions and events and learn on their own without input from their original coder or programmer means that there now is an objective world that exists outside of our mind and this world is one we cannot adequately conceive of through Kantian categories.

Here we turn back to Bostrom's doomsday argument on AI and malignant failures that outlined the ways in which supper-intelligent AI can go into a malignant mode of failure due to its overarching superiority over all other types of intelligence(s). Specifically, Bostrom postulated that if programmers make an error in their code or do not calculate all possible sets of actions for a given superintelligent application or machine, then the application can then turn to malignant future behavior which runs risk of bringing out hugely adverse outcomes. Even if an application is programmed to achieve a given outcome or goal, Bostrom demonstrates that the problem of perverse instantiation can lead to the program achieving the goal through other ways than originally programmed by the human because the superintelligent machine will not have our intrinsic biases and most importantly, it will not possess the same categories of our understanding–space and time. This really gets into the basic fact that I outlined above, in that AI may indeed operate outside of Kantian categories. Another very important dynamic can be observed in what Bostrom refers to as the infrastructure profusion problem. This practical projection entails that an intelligent machine or application can produce infrastructure profusion in which it will take and transform large portions of infrastructure that will benefit its own motives, interests, and goals. This process is said to accompany AI's ability to maximize its output according to a given outcome. These different examples are not all that Bostrom brings up in his noteworthy work, but they do nevertheless point to quite dangerous developments associated with AI.

Point Five

Mechanisms of AI are not the only mechanisms associated with this immensely profound and transformative technological development–the previous chapter identified numerous others which point to the real empirical fact that AI is a multifaceted phenomenon that spurs mechanistic activities and forces in different spheres of social reality. The ongoing digital revolution has probably been most influenced by the creation, presence, and continuous development of AI. Medical applications of these technical developments will only continue to increase and may become more profound in changing the course of illness in individuals' lives than the discovery of bacteria did just over a century ago. AI is progressively being applied in the medicinal field: preventive medicine, cancer screenings, among numerous other spheres such as ophthalmology.

The importance of these applications of AI in medicinal fields is that they present a data driven approach that quantifies processes which also went into the discovery of crucial phenomena such as bacteria which also were data driven, but were of a more qualitative character. For example, in one of the major discoveries that led to the identification of bacteria centuries ago, a Hungarian doctor named Ignaz Semmelweis carried out an experiment in the Austro-Hungarian empire in two Vienna maternity wards in 1846. One of the maternity wards had only students and male doctors while the other had only female midwives. He discovered that more women were dying as a result of child-bed fever in the all-male clinic when compared to the other clinic. It turned out that in the male clinic, doctors were frequently doing autopsies while the other clinic was not. From these observations, Semmelweis hypothesized that both doctors and their assistants in the all-male clinic were contracting bacterial particles form autopsies on their hands and then carrying these particles over to birth deliveries and giving the bacteria to the mothers (who then would die). As a remedy, the doctor recommended that staff wash their hands and use choline to kill germs. After a large number of repetitions and observations of contact tracing after what happened post-autopsies, the doctor realized that more babies survived their birth when the deliverers' hands were washed before delivering a baby.

In contemporary applications of AI to medical fields and practice, processes akin to Semmelweis' experiments in 1846 Vienna are taken to a completely new super-human level that would have been unimaginable for doctors in the early 20th, not to mention, 19th centuries. AI can scan peoples' arteries, bones, organs (including brains), and detect abnormalities, early tumor growth, plaque buildup, among other problems that are commonly experiences across human populations. The way in which AI applications can accomplish such feats is through data-driven medicinal inquiry–images, scans, and 3D processing from hundreds of thousands (and in some fields, millions) of past positive cases (i.e., in which an adversity was detected) are weighed against negative cases (of healthy subjects). Through immense computational power, a given AI application can then determine the likelihood that someone has one of the aforementioned problems in its early stages with great precision. In post-industrialized countries, these practices are already taking on mainstream adoption such as through Google's AI Healthcare system which can detect abnormalities on the level of a cell through probing human tissues.

The vast array of developments connected to AI have brought about a totally different set of social conditions than any previous point of human history because there now exists an objective world that operates outside of our cognitive awareness. Although Kant lived centuries before the emergence of such a configuration of human organization, the presence of an objective digital realm may contradict the Kantian claim that an objective world does not (and cannot) exist independent of our mind. Although this world (which I argue scholars in the future should refer to as a digital realm), is still in its early stages of development because it is limited to handheld devices, computers, and externally formatted methods of viewing (for individuals that use it), the aforementioned (chapter six) processes surrounding Neuralink and similar projects provide some preliminary evidence as to what the future may entail. This will heighten an already prominent tendency that Berdyaev identified begun with the rise of science, "Science and scientific foresight give man power and security, but they can also devastate his consciousness and sever him from reality" (Berdyaev 1937, 6).

In my view, the future will see greater autonomous sources of mechanistic forces operating and bringing about social outcomes. The expansion of the digital realm is indefinite, Moore's law, to the best of my knowledge, has not slowed down or receded (even if we consider the supply-chain problems that arose in 2021 and 2022 due to the COVID-19 pandemic). Since technological progress is so rapid and almost deterministic in its nature, there will be more computers, more smart devices, more digitalization of daily tasks and the general digitalization of socio-cultural practices. Eventually, there will arise an incremental phenomenon in which societies (beginning with the affluent, post-industrialized parts of the world) will gradually be a movement into a bio-synthetic network that will connect to humans to technological networks and systems. Throughout this historically transformative trajectory, the average time spent on technological devices will increase until a point that will render holding such devices in one's hand as infantile–this is precisely when some will begin encouraging new configurations of human existence. The interconnection between our brain, nervous system, and the digital realm will reach a point of no return–a point that will see a reversal of Kantian logic. The digital realm, AI, and computers, will instill their mental categories onto our mind, rather than our mind instilling mental categories on the world. When this happens, there likely will be a different, and at the point of writing this book, unimaginable form of contingency associated with social causality because we will not be the primary agents that articulate categories to investigate our world.

References

Berdyaev, Nikolai. 1937. *The Destiny of Man*. London: Geoffrey Bles.

Bostrom, Nick. 2014. *Superintelligence: Paths, Dangers, Strategies*. Oxford: Oxford University Press.

Kant, Immanuel. 1797. *Critique of Pure Reason. Trans. Norman Kemp Smith, 1997*. New York: Palgrave Macmillan.